ABOUT ISLAND PRESS

Island Press, a nonprofit organization, publishes, markets, and distributes the most advanced thinking on the conservation of our natural resources—books about soil, land, water, forests, wildlife, and hazardous and toxic wastes. These books are practical tools used by public officials, business and industry leaders, natural resource managers, and concerned citizens working to solve both local and global resource problems.

Founded in 1978, Island Press reorganized in 1984 to meet the increasing demand for substantive books on all resource-related issues. Island Press publishes and distributes under its own imprint and offers these services to other nonprofit organizations.

Support for Island Press is provided by the Geraldine R. Dodge Foundation, The Energy Foundation, The Charles Engelhard Foundation, The Ford Foundation, Glen Eagles Foundation, The George Gund Foundation, William and Flora Hewlett Foundation, The John D. and Catherine T. MacArthur Foundation, The Andrew W. Mellon Foundation, The Joyce Mertz-Gilmore Foundation, The New-Land Foundation, The J. N. Pew, Jr. Charitable Trust, Alida Rockefeller, The Rockefeller Brothers Fund, The Rockefeller Foundation, The Tides Foundation, and individual donors.

Ecosystem Health

Ecosystem Health

New Goals for
Environmental Management

EDITED BY

Robert Costanza
Bryan G. Norton
Benjamin D. Haskell

ISLAND PRESS
Washington, D.C. • Covelo, California

LIBRARY OF CONGRESS
CATALOGING-IN-PUBLICATION DATA

Ecosystem health : new goals for environmental
    management / edited by Robert Costanza, Bryan G.
    Norton, and Benjamin Hakell.
        p. m.
    Includes bibliographical references and index.
    ISBN 1-555963-141-4 (cloth). — ISBN 1-55963-140-6
    (paper)
    1. Human ecology—Philosopy. 2. Environmental
policy. 3. Human ecology—Moral and ethical aspects.
4. Ecology—Philosophy. I. Costanza, Robert. II.
Norton, Bryan G. III. Haskell, Benjamin D.
GT21.E26 1992
304.2—dc20   92-10531
                                                  CIP

Printed on recycled, acid-free paper

Manufactured in the United States of America

10 9 8 7 6 5 4 3 2 1

Contents

• PREFACE ix

• Introduction: What Is Ecosystem Health and Why Should We
      Worry About It?
      Benjamin D. Haskell, Bryan G. Norton,
      and Robert Costanza 3

• Part I: Philosophy and Ethics

• 1. A New Paradigm for Environmental Management
      Bryan G. Norton 23

• 2. Aldo Leopold's Metaphor
      J. Baird Callicott 42

• 3. Has Nature a Good of Its Own?
      Mark Sagoff 57

• 4. Toward an Open Future: Ignorance, Novelty, and
      Evolution
      Malte Faber, Reiner Manstetten, and John Proops 72

• 5. Environmental Existentialism
      Talbot Page 97

• 6. Environmental Therapeutic Nihilism
      Eugene Hargrove 124

• Part II: Science and Policy

• 7. Ecosystem Health and Ecological Theories
      David Ehrenfeld 135

• 8. What Is Clinical Ecology?
      David J. Rapport 144

• 9. Establishing Ecosystem Threshold Criteria
    David J. Schaeffer and David K. Cox   157

• 10. Alternative Models of Ecosystem Restoration
    Susan Power Bratton   170

• 11. Ecosystem Health and Trophic Flow Networks
    Robert E. Ulanowicz   190

• 12. Measures of Economic and Ecological Health
    Bruce Hannon   207

• 13. Ecological Integrity: Protecting Earth's Life
       Support Systems
    James R. Karr   223

• 14. Toward an Operational Definition of Ecosystem
       Health
    Robert Costanza   239

• INDEX   257

• CONTRIBUTORS   266

Preface

This book is significant because several environmental management agencies (especially EPA) are adopting a broader set of management goals that go beyond protecting human health to protecting ecosystem health. Although there is a need to better define and operationalize this concept, but the published literature offers little help. In the following pages, therefore, an outstanding group of ecologists, philosophers, economists, and others address this important topic at a most opportune time.

The book has several purposes: as a policy statement and research agenda for environmental management agency personnel, other policymakers, and the informed environmental public; as a textbook on "Ecosystem Health"; and as an academic text aimed at researchers in the field. Since the field is inherently multidisciplinary, the potential audience is multiplied. It should be of interest to economists, ecologists, philosophers, public policymakers, anthropologists, sociologists, and others.

We have tried to limit jargon and define terms as they appear so that they will be accessible to a broad audience. But since this is a book with many authors, the style varies considerably from chapter to chapter. We have not tried to force the diverse styles into a particular mold. Indeed, we think the many perspectives offered in this volume are representative of the pluralistic nature of the topic. The introductory chapter synthesizes these many perspectives and serves as a guide to the rest of the book.

Two meetings formed the basis for the material. First, a workshop was held at the Aspen Institute on Maryland's eastern shore on 20–23 October, 1990. The workshop brought together seventeen participants from academia, government, business, and private organizations representing a broad range of disciplines, including ecology, resource management, public policy, economics, and philosophy. The purpose of the workshop was to arrive at some consensus regarding the definition of ecosystem health. The second precursor was a symposium sponsored by the International Society for Environmental Ethics and presented at the American Association for the Advancement of Science meeting in Washington, D.C., in February 1991. The book is a compilation of papers presented at these two meetings, as well as some additional essays that were added to round out the collection.

ACKNOWLEDGMENTS

This project was supported by the Maryland Sea Grant Program (project title: Workshop on Ecosystem Health and Environmental Management; (contract no. 07-5-25126) principal investigators: R. Costanza and M. Sagoff); by the US EPA (contract no. 68-C9-0029); and by the Aspen Institute. Support is also gratefully acknowledged from the National Science Foundation (contract no. BBS-8619104), which enabled research leading to the meeting that produced papers for this volume. The views expressed are those of the authors, not necessarily those of any sponsoring organization.

Thanks are due especially to Chris D'Elia and Jack Greer of Maryland Sea Grant and Tom DeMoss of EPA for their encouragement and participation. Other participants in the Aspen workshop or the AAAS symposium who did not contribute a chapter to the book but made important contributions to the discussion include: Michael Deak, E. I. DuPont de Nemours & Co.; Fran Flanigan, Alliance for the Chesapeake Bay; Michael Haire, Maryland Department of the Environment; Larry Harris, University of Florida; Michael Hirschfield, Chesapeake Bay Foundation; Greene Jones, USEPA; Greg Michaels, ABT Associates; John Paul, USEPA; Henry Regier, University of Toronto; Charles Spooner, USEPA Chesapeake Bay Liaison Office; Laura Westra, International Society for Environmental Ethics; and Senator Gerald Winegrad, Maryland State Senate. Special thanks go to Joy Bartholomew of the Center for Policy Negotiation for ably facilitating the Aspen workshop. We would also like to thank Janis King for her help with copyediting and proofing and Jill Cowie for her patience.

Ecosystem Health

Introduction

What Is Ecosystem Health and Why Should We Worry About It?

BENJAMIN D. HASKELL, BRYAN G. NORTON,
AND ROBERT COSTANZA

This book addresses the lack of a clear conception of the term "health" in relation to ecosystems and analyzes the normative, conceptual, and biological issues surrounding the concept of ecosystem health. While "health" and "integrity" are used ubiquitously in long-range policy documents as goals of protection efforts, they have never been defined well enough to make them useful in practice. A clearer conception of the quality and health of natural environments will help bring our regulatory mandates in line with legislative ends. This book elaborates this policy issue to a broad audience in the environmental community.

Plainly, this concept cannot be defined or understood simply in biological or ethical or aesthetic or historical terms. Many approaches must be used in clarifying the goals of environmental protection. We have assembled a pluralistic, multidisciplinary collection of perspectives in this volume, covering a broad spectrum of ideas from philosophy, science, and management. Our intention is to provoke thought and debate—not to lay down dictums, but to move the debate forward toward practical principles.

DEFINING ECOSYSTEM HEALTH

The need to define the concept of ecosystem health in practical terms arises because of the failure of current economic paradigms to protect the natural environment, which is, after all, the foundation for economic systems. The ecosystem health paradigm proposed in this book uses a broad medical model, even while recognizing that the parallel between medicine and environmental protection does not always hold (see Norton 1991b). The unifying idea of management is nevertheless the idea of protecting and restoring health to ecological processes at all levels. The goal of this dynamic process is to protect the autonomous, self-integrative processes of nature as an essential element in a new ethic of sustainability.

Health is most easily defined negatively—that is, as the absence of disease. We prefer, however, a definition that states more positively the characteristics of healthy systems. The various disciplines—systems science, management and philosophy, and ethics—offer definitions, but these are not mutually exclusive and, in fact, tend to converge.

The notion of sustainability has become an important environmental policy supported by international organizations and a growing number of nations. Criteria for sustainability have been given a variety of formulations. Some emphasize sustainability of narrowly defined "economic" productivity over time. Others insist on ecologically based criteria. This book decisively adopts the latter approach and searches for an adequate formulation of the intuitive idea of a "healthy" system being valuable beyond benefits that can be computed simply in terms of consumptive preferences. In particular, the book adopts a generally hierarchical approach and defines ecosystem health as a characteristic of complex natural systems. Since fast-changing human cultures are embedded in larger-scale, slow-changing ecological systems, we must develop policies that allow human cultures to thrive without changing the life support functions, diversity, and complexity of ecological systems.

Environmental literature and the media are rife with examples of how humans are degrading the biotic and abiotic natural world (for example: destruction of species and ecosystems, global warming, toxic waste, acid rain). The rapid acceleration of technological growth associated with industrial and postindustrial societies poses, through cumulative impacts of many individual decisions, a risk to the health of normally slow-changing ecological systems. These negative impacts result from our use of the environment strictly for our needs with no regard for the sustainability of the global ecosystem. Mark Sagoff (Chapter 3) outlines three reasons why humans value the environment: instrumental, aesthetic, and moral. Until now, a narrow set of instrumental reasons for valuing the environment has dominated policy formulation at the expense of a more comprehensive view of instrumental value, or the aesthetic and moral reasons, leaving us with an imperiled biosphere. As Sagoff puts it: "Environments that are most productive economically are not necessarily those that have the most value aesthetically or ethically" (pers. comm.). Many of our nation's waterways for example, are unhealthy because we have used them to maximize our short-term, narrowly defined welfare by dumping our wastes into them, manipulating their natural characteristics, and using them as liquid highways.

We recognize that people will always value nature for different reasons and, therefore, a pluralistic approach is necessary to address this problem. It is clear that in fairness to the future of our culture and our species we must stop the deplorable destruction of natural processes and the dynamic creativity they embody. Whether this protection of ecosystem health is justified on the basis of an expanded view of instrumental value, or on the basis of aesthetic and moral criteria, might be argued. We think that a suitably comprehensive and long-term view of instrumental value—one that protects ecosystems' services by protecting the health and integrity of systems indefinitely—is sufficient to make the case and will support aesthetic and moral values as well. Moreover, an ecosystem health approach is needed because complex ecosystems must be assessed in ways that reflect their overall performance and not just the production of desired species. Single-species management has led to many disasters in the past, but before there can be true multicriteria management there must be an adequate definition of health.

The dictionary definition of health is: "1. the condition of being sound in mind, body, and spirit; 2. flourishing condition or well-being." While this definition is rather vague, it may be possible to construct a more rigorous and operational definition that is applicable to all complex systems at all levels of scale, including organisms, ecosystems, and economic systems. Such a general definition would guide our assessments of the stress on ecosystems and the amount of stress they can endure. Another purpose in defining ecosystem health is to provide direction to the process of ecological/economic integration. This definition must allow for the fact that ecological systems provide the foundation for human economic systems. Accordingly, the definition of ecosystem health is not meant to be an endpoint but rather one step in the ongoing process of understanding the relationship between economic and ecological systems.

Norton (1991a) suggests five axioms of ecological management that create a framework for defining health:

• The Axiom of Dynamism: Nature is more profoundly a set of processes than a collection of objects; all is in flux. Ecosystems develop and age over time.
• The Axiom of Relatedness: All processes are related to all other processes.
• The Axiom of Hierarchy: Processes are not related equally but unfold in systems within systems, which differ mainly regarding the temporal and spatial scale on which they are organized.
• The Axiom of Creativity: The autonomous processes of nature are creative and represent the basis for all biologically based productivity. The vehicle of that creativity is energy flowing through systems which in turn find stable contexts in larger systems, which provide sufficient stability to allow self-organization within them through repetition and duplication.
• The Axiom of Differential Fragility: Ecological systems, which form the context of all human activities, vary in the extent to which they can absorb and equilibrate human-caused disruptions in their autonomous processes.

The Chesapeake Bay, a well-known example of a cherished but heavily impacted ecosystem, is often cited as an example of an unhealthy system. The Chesapeake Bay has been the subject of much study and a plan for its restoration is under way. The unspoken goal of the effort is to "restore" the bay to its 1950 condition. This may be an unrealistic goal, however, since the bay has evolved since 1950 for both natural and cultural reasons. The restoration effort may be more successful with a more dynamic and flexible definition—one that allows for growth and development of an ecosystem including its human component.

MEASURING ECOSYSTEM HEALTH

Defining ecosystem health is a process involving the identification of important indicators of health (such as a species or a group of species), the identification of important endpoints of health (such as relative stability and creativity), and, finally, the identification of a healthy state incorporating our values. Robert Costanza illustrates this progression in Figure 1 of Chapter 14, where indicators and endpoints do not require much integration and are quantifiable with a fairly high degree of precision. Measures of a "healthy state" are less precise but are much more comprehensive and relevant and require integration and modeling.

Before health can be measured, we need to identify the relevant indicators, endpoints, and parameters (with acceptable ranges) that we are going to use in assessing the health of a particular ecosystem. Over the past forty years many such variables and parameters have emerged in the ecological literature, but there are several problems with identifying relevant endpoints and indicators. First, each ecosystem has its own set of indicators and endpoints; therefore each ecosystem must be assessed separately. Second, the process of succession presents a problem: if conditions change, a new ecosystem that is better adapted to the new conditions will replace the prior system. Consequently, our indicators and variables must be sufficiently dynamic to change accordingly. Third, each scientist evaluating the ecosystem will choose different variables depending on his or her specific interest and expertise.

Historically, the health of an ecosystem has been measured using indices of a particular species or component. Such indices are inadequate because they do not reflect the complexity of ecosystems. From a managerial or political standpoint, however, these single species indices are sometimes useful in communicating complex information to resource managers, politicians, and the public. New indices have been developed that better reflect the complexity of the ecosystem and are being used for management purposes. For example, about forty states are using the Index of

Biotic Integrity (a sum of twelve characteristics of the fish com-/ munity) developed by James Karr (Karr 1981; Karr et al. 1986; Karr 1991; Chapter 13 of this volume) to evaluate stream water quality. Robert Ulanowicz (1980; Chapter 11) developed the Index of Network Ascendency which incorporates four important characteristics of ecosystems: species richness, niche specialization, developed cycling and feedback, and overall activity. In evaluating the risks to the health of the nation's estuaries, EPA and NOAA have developed a stress index based on a compilation of potential stresses such as pollutant discharge and sediment runoff. Although indices are potentially useful tools, they must reflect the underlying complexity of the ecosystem and the conceptual and scientific basis of health.

Schaeffer et al. (1988) suggests some guidelines to follow in our assessment of ecosystem health:

1. Health should not depend on criteria based on the presence, absence, or condition of a single species.

2. Health should not depend on a census or even inventory of large numbers of species.

3. Health should reflect our knowledge of normal succession or expected sequential changes that occur naturally in ecosystems.

4. While the optimal health measures should be single-valued (monotonic) and vary in a systematic and discernible manner, ecosystem health does not have to be measured as a single number. Single numbers compress a large number of dimensions (one for each type of item) to a point that geometrically has zero dimensions.

5. Health measures should have a defined range.

6. Health criteria should be responsive to change in data values but should not show discontinuities even when values change over several decades.

7. Health measures should have known statistical properties, if these are relevant.

8. Criteria for health assessment must be related and hierarchically appropriate for use in ecosystems.

9. Health measures should be dimensionless or share a common dimension.

10. Health measures should be insensitive to the number of observations, given some minimum number of observations.

Another major problem in defining the health of an ecosystem is choosing the appropriate scale (Chapters 1 and 3). What ecosystem are we managing and what level are we focusing on? The scale that is chosen must provide the right balance between resolution and predictability. In our attempts to manage the Chesapeake Bay ecosystem, for example, do we focus on the estuary itself or do we focus on the whole 64,000-square-mile watershed?

AN OPERATIONAL DEFINITION

The workshop participants arrived at a working definition of ecosystem health that incorporates most of the considerations just listed. It defines health in terms of four major characteristics applicable to any complex system: *sustainability*, which is a function of *activity*, *organization*, and *resilience*. Thus they concluded:

> An ecological system is healthy and free from "distress syndrome" if it is stable and sustainable—that is, if it is active and maintains its organization and autonomy over time and is resilient to stress.

This definition is applicable to all complex systems and allows for the fact that ecosystems are growing and developing in response to both natural and cultural influences. The definition employs several concepts that are defined in Table 1 of Chapter 14. A key concept in this definition is sustainability, which implies that the system can maintain its structure and function over time. Using this definition, a diseased system is one that is not sustainable, that will eventually cease to exist. "Distress syndrome" refers to the irreversible process of system breakdown leading to collapse. This definition is consistent with Karr's definition from 1986 (Karr 1986; chapter 14, this volume).

RESEARCH AGENDA AND RECOMMENDATIONS

The process of defining ecosystem health has begun. What we need now is a long-term strategy for the study and improvement of ecosystem health. We propose to use the practice of human and animal medicine as the model for the practice of "ecological medicine." Medical practitioners have a defined process for assessing health with the help of a compendium of known diseases and a huge body of reference data on the "standard human." Doctors rely extensively on their own experience and the experience of others to diagnose the problem correctly. The assessment of health in medicine follows this sequence:

1. Identify symptoms.
2. Identify and measure vital signs.
3. Make a provisional diagnosis.
4. Conduct tests to verify the diagnosis.
5. Make a prognosis.
6. Prescribe a treatment.

This model of health assessment is applicable to ecosystems, but ecologists do not have at their disposal a compendium of known diseases or stresses with associated symptoms and signs. Although it is therefore impossible to make an accurate diagnosis and prescribe the appropriate treatment for an ailing ecosystem, ecologists have neverthless started to categorize certain stresses and record their effects (symptoms) on the ecosystem. Rapport et al. (1985), for example, have reviewed these stresses and their symptoms (Tables 1 and 2).

There is another sense in which ecosystem treatment lags behind medicine: whereas physicians can usually concentrate their efforts on the health of a given .i.patient;, ecosystems exist on many levels and can be described on many scales. More research on the rational choice of a scale from which to manage natural systems seems essential, especially in those areas where public goals are not articulated clearly (see Chapter 1). Scientific principles for the choice of scales and development of practical assessments may only proceed in those cases in which a consensus of public goals can be developed. If public goals are consensually accepted, it is already possible to formulate diagnostic tests for ecosystem stress (Norton, Ulanowicz, and Haskell 1991).

TABLE 1. *Symptoms of Ecosystem Distress*

	Response Variable					
Type of stress	Nutrient pool	Primary productivity	Size distribution	Species diversity	System retrogression[a]	Source
Harvesting of renewable resources						
Opportunistic fishing	*	*	-	-	+	Regier & Loftus (1972)
Commercial fishery; Great Slave Lake	*	*	-	-	*	Keleher 1972)
Deforestation; northern hardwoods[b]	-	-	-	-	+	Likens et al. (1978)
Pollutant discharges						
Radiation; oak/pine forest	*	*	-	-	+	Woodwell 1967)
Nutrient; lakes	+	+	-	*	+	Colby et al. (1972)
Toxins; rivers	*	*	-	-	+	Cairns & Dickson 1977)
SO_2; boreal forest	-	-§	-	-	+	Gorham & Gordon 1960)
SO_2; coniferous forest; Poland	*	-	-	-	+	Wolak 1979)
Oxidants; forest, San Bernadino Mts., Calif.	*	-	-	-	+	Miller et al. (1963); Miller 1973)

Note: Signs (+ or -) indicate direction of change compared with normal functioning of relatively unstressed systems.

* Response was not monitored.

§ Inferred from decrease in vegetation cover and nutrient pools.

[a]The symbol + indicates that retrogression has occurred.

[b]Although the Hubbard Brook clear-cut was not designed to simulate a commercial harvest of the forest, "it soon became apparent that the results could provide significant insights into environmental questions posed by commercial cuttings in northern hardwood forests" (Likens et al. 1978:492).

[c]Possibly acid rain; major effects noted so far are on spruce and beech.
Source: Rapport et al. (1985)

TABLE 1. Continued

| | Response Variable | | | | | |
Type of stress	Nutrient pool	Primary productivity	Size distribution	Species diversity	System retrogression[a]	Source
Arsenic; boreal forest, Yellowknife, NWT	*	-	-	-	+	Hutchinson et al. (1982)
General air pollution; Central European forests[c]	-	-	*	-	+	Ulrich et al. (1980)
Physical restructuring						
Coal surface mining; Ohio	-	-	-	-	+	Riley 1977)
Shoreworks; Italian subalpine lakes	*	*	-	-	+	Grimaldi & Numann 1972)
Toxic mine tailings; NWT	-	-	-	-	+	Kuja & Hutchinson (1979)
Pipeline construction; Arctic	+	-	-	-	+	Van Cleve (1972)
China clay waste; Cornwall, U.K.	-	-	*	-	+	Bradshaw et al. (1972)
Introduction of exotics: fish species	*	*	-	*	+	Regier (1973)
Extreme natural events						
Hurricane; coral reef	*	*	*	-	+	Woodley et al. (1981)
Earthquake; tropical forest	*	-	-	-	+	Garwood et al. (1979)
Multiple causes						
Baltic Sea	+	+	*	*	+	Leppakosi (1980)
Great Lakes	+	+	-	*	+	Regier & Hartman (1973)

TABLE 2. *Characteristic Response of Ecosystems to Stress*

Type of Stress	Nutrient pool	Primary productivity	Size distribution	Species diversity	System retrogression
Harvesting of renewable resources					
Aquatic	*	*	-	-	+
Terrestrial	-	-	-	-	+
Pollutant discharges					
Aquatic	+	+	-	-	+
Terrestrial	-	-	-	-	+
Physical restructuring					
Aquatic	*	*	-	-	+
Terrestrial	-	-	-	-	+
Introduction of exotics					
Aquatic	*	*	*	-	+
Terrestrial	*	*	*	*	+
Extreme natural events					
Aquatic	*	*	-	-	+
Terrestrial	-	-	-	-	+

Note: Signs (+ or -) indicate direction of change compared with normal functioning of relatively unstressed systems.
* Characteristic response not sufficiently determined.
Source: Rapport et al. (1985).

After a list of stresses and symptoms is compiled, the next step is to develop a series of quick diagnostic tests that will enable the ecologist to detect stress early (before the ecosystem begins to retrogress) and recommend remedial action. Once a compendium of diseases has been compiled and diagnostic tools have been developed, we can then practice "preventive medicine." Problems such as the overenrichment problem in the Chesapeake Bay can be avoided before they occur.

The Chesapeake Bay may serve as a model for the practice of ecological medicine. There are indications that the bay is entering a state of hypereutrophication. The signs and symptoms of hypereutrophication are shorter food chains with a lower degree of organization, lower diversity, and higher productivity—mostly at the bottom of the food chain. An early study of the bay's problems identified nutrient overenrichment as the principle cause of the bay's declining health. A multi-state management program was initiated and steps were taken to stem the flow of nutrients. The bay management program has progressed through all six stages in the health assessment process outlined above.

Lacking a compendium of diseases and diagnostic tools for assessing ecosystem health, we recommend further development of network analysis and simulation modeling. Network analysis (one example of which is Ulanowicz's ascendency index) allows for integrated, quantitative, hierarchical treatment of all complex systems including ecosystems. Measures like ascendency estimate not only how many different species there are in a system but, more important, how these species are organized. Simulation modeling of human impacts allows us to predict ecosystem response as a result of various site-specific management alternatives and natural changes. Development of this capability is essential not only for regional ecosystem management but also for modeling regional and global ecosystem response to regional and global climate change, sea level rise resulting from atmospheric CO_2 enrichment, acid precipitation, toxic waste dumping, and a host of other potential impacts.

SCIENTIFIC THEORY OR ENVIRONMENTAL POLICY?

A key question remains open: whether, and to what extent, the idea of ecosystem health should be allied most closely with scientific theories or with policy goals. Robert Ulanowicz seeks to embed the concept of ecosystem health in a theory of ecosystem development, expecting that health will correlate with quantifiable indices associated with highly developed ecological systems. James Karr and David Ehrenfeld, however, caution against too tight a connection between the intuitive concept of health and a single, specific ecological theory. Mark Sagoff argues that ecosystem health must be addressed in a policy framework because it involves identifying values.

We believe there exists considerable basis for expanding consensus if the concept of health is given its primary identity as a policy concept: a general guide to formulation of policy goals. Then it will be possible to explore a variety of scientific measures for their usefulness in providing descriptive criteria of use in formulating environmental policies. We accept J. Baird Callicott's historical account of the concept of ecosystem health (Chapter 2)—it emerged in the mature thought of Aldo Leopold as a crucial bridge between hands-on technical management and the ongoing task of formulating management goals. Following Leopold, we advocate an open mind and an experimental spirit in both the formulation of goals and the development of management tools to pursue them.

One important area of disagreement concerns the nature of ecological theory's contribution to environmental management. Sagoff (1988) has summarized arguments that theoretical systems analysis has contributed little to our understanding of environmental management. We accept the points of Ehrenfeld and Karr, recognizing the pitfalls of developing the definition of ecosystem health in rigid connection with a specific theory of ecosystem structure and function. We therefore adopt a *minimalist systems approach*. That is, we believe that some form of general systems analysis should provide the basic approach to conceptualizing interactions between human cultures and more natural systems that provide their context, but the definition of ecosystem health should not be linked too closely to specific empirical hypotheses. We interpret this more as a choice of paradigm—a constellation of conceptualizations, basic assumptions, and methods. While some authors in this volume adopt a rather comprehensive systems theory (Ulanowicz, Norton, and Costanza), this need not be seen as embodying specific, empirically determined system characteristics, such as the idea that ecological systems progress toward a single, characteristic climax state. We do not endorse full-blown organicism—the notion that ecological systems unfold according to formal cause (telos) or that ecosystems behave like organisms. We use organicism, rather, as simply a metaphor. In fact, open, nonliving systems—especially dissipative structures—provide a more robust conceptual analog than do either mechanism or organicism. Dissipative structures, from chemical clocks to complex ecosystems, maintain their system and structure through interchanges of energy with their environment. And while it is often useful to note the similarities between organized systems and organisms, use of these metaphors and analogies need not imply a literal assertion that on any given level ecosystems are alive.

This simple formalism of interpreting ecosystems as hierarchically organized systems begs as few empirical questions as possible, and leads us to focus on hierarchy, scale, and multiple levels of organization. It also leads us to be wary of haphazard aggregation of measures across levels. At this point we favor a multidimensional, open approach to systems. Arguments by the New Physicists suggest that there can be no consistent and complete description of the world (Prigogine and Stengers 1984). This conclusion squares with antireductionistic arguments in the philosophy of science, as well as incompleteness results in logic, and the philosophy of mathematics (Quine 1951; Goëdel 1931). That is, the idea of paradigms, which accepts that every style of description rests inevitably on posited, constitutive assumptions, is an unavoidable consequence in both philosophy and physical theory. If the assumptions embodying the paradigms of various disciplines cannot be made commensurate, reduction across levels will be impossible.

We are therefore uncertain about the possibility of developing "filters"—formulas that relate measured parameters on one level to those on another (Allen and Starr 1982). It may be that the various scales of the system are driven by very different dynamics (Norton, Ulanowicz, and Haskell 1991) and that the measures best used to describe those multiple levels are incommensurate. It remains an open question, therefore, to what extent aggregation can be accomplished in the description of ecosystem characteristics. Minimal systems theory—ecology without teleology—simply emphasizes multiple levels, views these in hierarchical fashion, and pays close attention to the scale of the dynamic processes we study and manage. More specific attributions of characteristics of systems should be considered a matter for experimentation and empirical determination.

We take it as unquestioned that it would be helpful to have an integrated theory of environmental management, one that provides theoretical justification for the relationships we perceive among the different elements—economic, ecological, and political—of environmental policy and management. As Hannon and Ulanowicz (Chapters 11 and 12) argue, there must be some theoretical explanation for our choice of health indicators. Our position is that a minimalist systems approach, guided by a broad analogy to medical and veterinary science, provides the right mix of flexibility and theoretical articulation. At least some theory is necessary if we are to explain why the measures we choose to indicate states of ecosystem health are not arbitrary. But that theory should be flexible enough to allow experimentation with a broad range of measures and theories for understanding ecosystems and with a variety of hypotheses regarding the goals we should seek in management.

CONCLUSION

We are advocating an adaptive, experimental approach to environmental management. What is distinctive about this approach is a recognition that environmental management is a prescriptive science, and that environmental goals shape science as well as vice versa. In this sense, social values—what we mean by a healthy system in which to live—are also open to experimentation. The central goal of this book is to further the discussion of this idea and to foster public debate and scientific debate regarding the proper goals for environmental management.

References

Allen, T. F. H., and T. B. Starr. 1982. *Hierarchy Theory.* Chicago: University of Chicago Press.
Bradshaw, A. D., W. S. Dancer, J. S. Handley, and J. C. Sheldon. 1975. "The Biology of Land Vegetation and the Reclamation of the China Clay Wastes in Cornwall." *British Ecological Society Symposium* 15:363–384.
Cairns, J., Jr., and K. L. Dickson. 1977. Pages 303–364 in J. Cairns, Jr., K. L. Dickson, and E. E. Herricks (eds.), *Recovery and Restoration of Damaged Ecosystems.* Charlottsville: University of Virginia Press.
Colby, P. J., G. R. Spangler, D. A. Hurley, and A. M. McCombie. 1972. "Effects of Eutrophication on Salmonid Communities in Oligotrophic Lakes." *Journal of the Fisheries Research Board of Canada* 29:975–983.
Costanza, R., F. H. Sklar, and M. L. White. 1990. "Modeling Coastal Landscape Dynamics." *BioScience* 40:91–107.
Garwood, N. L., D. P. Janos, and N. Brokan. 1979. "Earthquake-Caused Landslides: A Major Disturbance to Tropical Rainforests." *Science* 205:997–999.
Goëdel, K. 1931. "Über Formal Unendscheidbare Sätze der Principia Mathematica Und Verwandter Systeme." *Monatschrift für Mathematik und Physik* 38:173–198.

Gorham, E., and A. G. Gordon. 1960. "Some Effects of Smelter Pollution Northeast of Falconbridge, Ontario." *Canadian Journal of Botany* 38:307–312.

Grimaldi, E., and W. Numann. 1972. "The Future of Salmonid Communities in the European Subalpine Lakes." *Journal of the Fisheries Research Board of Canada* 29:931–936.

Hutchinson, T. C., S. Aufreiter, and R. V. N. Hancock. 1982. "Arsenic Pollution in the Yellowknife Area from Gold Smelter Activities." *Journal of Radioanalytical Chemistry* 71:59–73.

Karr, J. R. 1981. "Assessment of Biotic Integrity Using Fish Communities." *Fisheries* 6:21–27.

_____. 1991. "Biological Integrity: A Long-Neglected Aspect of Water Resource Management." *Ecological Applications* 1:1.

Karr, J. R., K. D. Fausch, P. L. Angermeier, P. R. Yant, and I. G. Schlosser. 1986. *Assessing Biological Integrity in Running Waters: A Method and Its Rationale.* Champaign: Illinois Natural History Survey, Special Publication 5.

Keleher, J. J. 1972. "Great Slave Lake: Effects of Exploitation on the Salmonid Community." *Journal of the Fisheries Research Board of Canada* 29:741–753.

Kuja, A. L., and T. C. Hutchinson. 1979. "The Use of Native Species in Mine Tailings Revegetation." Pages 207-221 in *Proceedings of Canadian Land Reclamation Association 4th Annual Meeting*, Regina, August 13–15.

Levins, R. 1966. "The Strategy of Model Building in Population Ecology." *American Scientist* 54(4):421–431.

Leppakoski, E. 1980. "Man's Impact on the Baltic Ecosystem." *Ambio* 9:174–181.

Likens, G. E., F. H. Bormann, R. S. Pierde, and W. A. Reiners. 1978. "Recovery of a Deforested Ecosystem." *Science* 199:492–496.

May, R. M. 1973. *Stability and Complexity in Model Ecosystems.* Princeton: Princeton University Press.

Miller, P. R. 1973. "Oxidant Induced Community Changes in a Mixed Conifer Forest." In J. A. Naegala (ed.), *Air Pollution Damage to Vegetation.* Washington: American Chemical Society.

Miller, P. R., J. R. Parmeter, O. C. Taylor, and E. A. Cardiff. 1963. "Ozone Injury to Foliage of *Pinus Ponderosa*." *Phytopathology* 53:1072–1076.

Norton, B. G. 1991a. *Toward Unity Among Environmentalists.* New York: Oxford University Press.

_____. 1991b. "Ecological Health and Sustainable Resource Management." In R. Costanza (ed.), *Ecological Economics: The Science and Management of Sustainability.* New York: Columbia University Press.

Norton, B. G., R. E. Ulanowicz, and B. D. Haskell. 1991. *Scale and Environmental Policy Goals*. Report to the EPA, Office of Policy, Planning, and Evaluation. Washington: EPA.

Pimm, S. L. 1984. "The Complexity and Stability of Ecosystems." *Nature* 307:321–326.

Prigogine, I., and I. Stengers. 1984. *Order Out of Chaos: Man's New Dialogue with Nature*. New York: Bantam.

Quine, W. V. O. 1951. "Two Dogmas of Empiricism." *Philosophical Review* 60:20–43.

Rapport, D. J. 1989. "What Constitutes Ecosystem Health?" *Perspectives in Biology and Medicine* . 33:120–132.

Rapport, D. J., H. A. Regier, and T. C. Hutchinson. 1985. "Ecosystem Behavior Under Stress." *American Naturalist* 125:617–640.

Regier, H. A. 1973. "Sequence of Exploitation of Stocks in Multispecies Fisheries in the Laurentian Lower Great Lake." *Journal of the Fisheries Research Board of Canada* 30:1992–1999.

Regier, H. A., and W. L. Hartman. 1973. "Lake Erie's Fish Community: 150 Years of Cultural Stresses." *Science* 180:1240–1255.

Riley, C. 1977. "Ecosystem Development on Coal Surface-Mined Lands, 1918-75." In J. Cairns, Jr., K. L. Dickson, and E. E. Herricks (eds.), *Recovery and Restoration of Damaged Ecosystems*. Charlottsville: University of Virginia Press.

Sagoff, M. 1988. "Ethics, Ecology and the Environment: Integrating Science and the Law." *Tennessee Law Review* 56(1):77–229.

Schaeffer, D. J., E. E. Herricks, and H. W. Kerster. 1988. "Ecosystem Health: 1. Measuring Ecosystem Health." *Environmental Management*. 12:445–455.

Ulanowicz, R. E. 1980 "An Hypothesis on the Development of Natural Communities." *Journal of Theoretical Biology* 85:223–245.

———. 1986 *Growth and Development: Ecosystems Phenomenology*. New York: Springer-Verlag.

Ulrich, B., R. Mayer, and P. K. Kharna. 1980. "Chemical Changes Due to Acid Precipitation in Loess-Derived Soil in Central Europe." *Soil Science* 130:193–199.

Van Cleve, K. 1972. "Revegetation Of Disturbed Tundra and Taiga Surfaces by Introduced and Native Plant Species." *Proceedings of Alaska Science Conference* 23:114–115.

Wolak, J. 1979. "Reaction des Ecosystemes; À la Pollution Subnecrotique." *Symposium On The Effects Of Airborne Pollution On Vegetation*. Warsaw:

Food and Agricultural Organization, Economic Commission for Europe, Joint Agricultural Timber Division.

Woodley, J. D., E. A. Chornesky, P. A. Clifford, et al. 1981. "Hurricane Allen's Impact on Jamaican Coral Reefs." *Science* 2114:749–755.

Woodwell, G. W. 1967. "Radiation and the Patterns of Nature." *Science* 156:461–470.

Part I

Philosophy and Ethics

1

A New Paradigm for Environmental Management

Bryan G. Norton

We can begin from the premise that our basic approach to environmental management must be rethought, a process that has already begun in earnest but remains in a state of considerable confusion. Indeed, the rethinking has progressed through the first stage—the critical and negative stage—as there now exist several cogent, convincing, and destructive critiques of the old approaches to management. These critiques reiterate the central insight of Aldo Leopold's land ethic: the old approach, which is characterized by reductionism, atomism, and economic determinism, is an inadequate general yardstick for measuring environmental gains and losses (Callicott 1989).[1] But the task of developing a new theory of environmental management, one that is more comprehensive, is far from complete. Theorists are proposing no less than construction of a new paradigm to guide environmental thought and action (Toulmin 1972).[2]

Changes on this profound level represent a radical departure in the way the world is described, and such changes normally gain explicit expression only slowly. An explicit search for a new paradigm of management is a conscious effort to "grow" new concepts more rapidly, as if in a greenhouse. Modern technology, and its efficiency is changing the entire biosphere in undesirable as well as desirable ways, has quite literally turned up the heat: our ability to change the world outruns our ability to understand the full ramifications of those changes. When the effects of our actions change the world too rapidly for natural linguistic change to keep up, we can catch up only if we can consciously construct a new, and more adaptive, set of terms and concepts (Toulmin 1972; Norton 1977).

The development of a new paradigm involves, by definition, the creation of a new constellation of axioms and concepts, an alternative set of assumptions, a new method. The paradigmatic approach, therefore, does not perceive the problem simply as one of adding a more holistic criterion of health to the set of concepts that constitute the current, reductionistic paradigm. Adopting a new paradigm represents a more radical departure; it is to interpret the world in a new format, a format that is given shape and structure by the development of new concepts and a vocabulary to express them.

DEVELOPING THE NEW PARADIGM

It is first necessary to explore the idea that there is an obligation to protect the health and integrity of ecological systems. Although some philosophers have insisted that this obligation is owed directly to these systems (Callicott 1989), I avoid that formulation here, basing the argument, instead, on obligations to future generations of humans. This argument recognizes the crucial role of creative, self-organizing systems in supporting human economic, recreational, aesthetic, and spiritual values.[3] Because self-organizing systems maintain a degree of stable functioning across time, they provide a sufficiently stable context to which human individuals and cultures can adapt their practices. Let us, therefore, posit the following moral principle of sustainability: No generation has a right to destabilize the self-organizing systems that provide the context for all human activity (Norton 1989; Rawls 1971).

We shall return to the difficult concepts of systemic creativity and self-organization later in this chapter. Here, we assume these problematic concepts and use them to develop a set of axioms and definitions that give general shape to the new paradigm of environmental management. Given working definitions of autonomy and self-organization, we can define sustainability as follows: Sustainability is a relationship between dynamic human economic systems and larger, dynamic, but normally slower-changing ecological systems, such that human life can continue indefinitely, human individuals can flourish, and human cultures can develop—but also a relationship in which the effects of human activities remain within bounds so as not to destroy the health and integrity of self-organizing systems that provide the environmental context for these activities.

The concepts of health and integrity are here understood as concepts in normative science. Like medicine, both conservation biology and environmental management employ science in the service of valued goals. Ecology is the "fusion point" of environmental science and management (Leopold 1939), and we can follow Faber, Manstetten, and Proops (Chapter 4 of this volume), who note that "ecology" combines, etymologically, the Greek ideas of "house" and "logos," the latter of which they translate as concept/structure. Therefore, they define ecology as "the science of the principles of the self-organization of nature." Thus defined, it is part of the very specification of environmental management that its subject matter is self-organizing. These definitions ensure that ecology, not economics, will be the science that provides the basic organization for the new management paradigm because economics is understood as one type of ecological activity, one that takes place within the economic system of one (however dominant) species.

Our task here is to build on this approach to management by providing more elements of the ecological paradigm for environmental management. To that end, consider these five axioms of ecological management:

• The Axiom of Dynamism: Nature is more profoundly a set of processes than a collection of objects; all is in flux.
• The Axiom of Relatedness: All processes are related to all other processes.
• The Axiom of Hierarchy: Processes are not related equally but unfold in systems within systems, which differ mainly regarding the temporal and spatial scale on which they are organized.
• The Axiom of Creativity: The processes of nature are self-organizing, and all other forms of creativity depend on them. The

vehicle of that creativity is energy flowing through systems that generates complexity of organization through repetition and duplication.

• The Axiom of Differential Fragility: Ecological systems, which form the context of all human activities, vary in the extent to which they can absorb and equilibrate human-caused disruptions in their creative processes.

These axioms function, in practice, in conjunction with normative definitions of ecosystem health and integrity. As Aldo Leopold (1949) noted in his comparison of succession after the plow in Kentucky and in the American Southwest, we require two related concepts to describe ecosystem well-being. In both cases, the integrity of the system was compromised—including loss of total diversity and invasion by exotics. The difference, Leopold noted, was that bluegrass represented a new stable point capable of maintaining itself. This system lacked the integrity of systems, like Rio Gavilan, that maintained their "aboriginal health" (including their historical mix of species) but maintained a "healthy" equilibrium nonetheless (Meine 1988). Let us say that a system is healthy if it maintains its complexity and capacity for self-organization. An ecological system has maintained its integrity—a stronger concept that includes the conditions of health—if it retains (1) the total diversity of the system—the sum total of the species and associations that have held sway historically—and (2) the systematic organization which maintains that diversity, including, especially, the system's multiple layers of complexity through time (Norton 1990).[4] An agricultural community, for example, can be healthy, even though many of its native species have been replaced with cultivated strains of exotic species, provided that the overall system maintains sufficient complexity to protect its self-organizing qualities.

Within this dynamic, contextualist paradigm we can understand the centrality of the goal of protecting biological complexity, which has become the keystone of the new theory of environmental management. Complexity is directly related to self-organization, and these characteristics are the essence of ecosystem health and integrity. The obligation to protect biodiversity can now be formulated as above all an obligation to protect biological integrity to the extent possible and to protect health where past actions have already destroyed the integrity of large systems.

Proponents of the new approach to environmental management are united by two beliefs: First, there must be an ecological component in the definition of sustainability; second, this ecological component will be stated as a requirement that human activities protect the complex processes of nature across generations. We call this theory "contextualism," because it implies that all human activities must be understood in the larger context of self-organizing systems.

SELF-ORGANIZING SYSTEMS

The problem of parts and wholes has intrigued philosophers since philosophy emerged from mythopoetic discourse. Solutions to this problem have had a deep impact on our culture's theories, behavior, and collective psyche. Recent developments in quantum theory, in information/systems theory, and in the physical sciences more generally suggest that the two main competitors among traditional solutions—atomism and more-or-less literal forms of organicism—are both inadequate. Newtonian dynamics, which was formulated in an atomistic and reductionistic paradigm that provided the primary method of the Renaissance worldview, emphasized analysis and, by abstraction, treats all true "individuals" as self-sufficient and independent (Earley 1990). This paradigm, when applied in any discipline from atomic physics to economics, abstracts the individual from relations to its context. Objects are reduced, whenever possible, to physical forces, and interactions among them are treated deterministically. The model of causation among natural objects assumed in this paradigm is deterministic and reversible—mechanistic. Even in physics (indeed, especially in physics), the old conceptual framework is recognized to be inadequate. (See Prigogene and Stengers 1984; Koestler 1967; Ulanowicz 1986.)

But the alternative offered by traditional organicists has proved no less problematic. Literal applications of the Aristotelian idea of growth and development according to a prior plan introduce teleology and the need to posit an intentional being that guides the organization of nature. Besides introducing difficult and untestable metaphysical elements into scientific discourse, these ideas run counter to the widely accepted model of Darwinian evolution, which posits random variation sorted by a blind tendency of the fittest to survive. The teleological approach seems to require, in addition to deterministic causation among individual elements, some form of "top-down" causation according to an intentional agency, which to reductionists is regarded as mystical—causation without mechanism. Some advocates of full-blown organicism are willing to accept these metaphysical implications; others, less mystically inclined, endorse organicism only as a shift in governing metaphors, not as an assertion of literal truth.

Even in its metaphorical interpretation, however, traditional organicism has a fatal flaw: in order to consummate the analogy to a self-directing organism, a coherent organism must reify some level of the system as "whole." This tendency is present, for example, in James Lovelock's Gaia theory, where it is assumed that the biosphere represents the organism level of systematic organization (Lovelock 1977, 1988; Joseph 1990).[5] But this approach simply transfers the assumption of independence that is applied to all objects in the reductionistic paradigm to an arbitrarily drawn boundary within the continuum of embedded systems of nature. By cutting off the systemic method at the level of the biosphere, Lovelock invites the criticism of mysticism: if one reifies one level of the hierarchy of systems *and* insists that the larger system shapes its subsystems to its own "purposes," it is difficult to avoid these charges.

While models such as Lovelock's may prove useful for some purposes, systems theory in its most general and robust form treats systems as open, meaning that they maintain their complex form and functions through continuous exchanges of energies and material with their environment (Koestler 1967). Not only does an open approach to systems theory defuse objections of mysticism, it offers also a more general framework of analysis in which information passed across levels of the system is interpreted nondeterministically in terms of probabilities, rather than as "intentions" or "purposes" (Ulanowicz 1986). In this general form, systems theory is hierarchical and open-ended, and causation flows both ways in the system, but top-down causation is not intentional in the narrower, teleological sense usually associated with mystical versions of vitalism and organicism. Because the proposed paradigm posits dynamic and open, rather than closed, systems, it reifies no level as The Whole and avoids the temptation to attribute intentional construction to whole ecosystems. Self-organization can be described phenomenologically with no metaphysical commitments to organicism or vitalism. (See Prigogene and Stengers 1984; Ulanowicz 1986.) The mathematics of self-organization, that is, is neutral with respect to teleology or intention on the part of the system. The new paradigm of ecological management can be based on common features of self-organizing and self-regulating systems whether organic or inorganic; living systems are a subset of self-organizing systems.

I have suggested that neither atomism nor holism is completely true. It is just as accurate to say that *both* are true, but they are true only within bounds. Both provide useful approximations of the truth considered from one perspective, on one scale, of the system. The processes of nature are, therefore, modeled as open, hierarchical systems that can be studied from a variety of perspectives and characterized using varied spatial and temporal scales (Allen and Starr 1982; O'Neill et al. 1986). Essential to this hierarchical structuring is the recognition of "membranes"—boundaries that separate systems and further subdivide the systems into subsystems.

Every description of the observable world must include an observer; this we have learned from quantum theory and relativity theory. Choices regarding the perspective and scale of the observer represent choices to create an abstract and less complete model than is represented in "total" reality. The essence of the new, nonreductionistic paradigm is to reject the assumption, implicit in the Renaissance worldview and in Newtonian mechanics, that the observer can adopt a neutral viewpoint from outside the system. The abstraction of reductionistic analysis is to remove the observer from nature in order to create deterministic models; models that place the observer in nature, on the other hand, cannot be strictly deterministic.

If both reductionism and holism are true, but on different levels, it follows that the structure of the observed world must be knowable on more than one level. Once one posits the observed world as hierarchically structured and, hence, fundamentally complex, one can see reductionism and systems analysis as applying appropriately to different levels of this complexity: *analysis* presupposes a system of independent monads and applies on the microlevel; *synthesis*, which is holistic and presupposes its objects are interrelated, applies in a larger system (Koestler 1967).

The key to a new resolution of the problem of parts and wholes will be the development of a method of understanding that is

powerful enough to guide policy but recognizes the role of creativity and innovation within an open system. (See Chapter 4 of
this volume and Boulding 1991.) The new paradigm of environmental management, which assumes hierarchical complexity,
adopts the self-organizing character of ecological systems as its
centerpiece. When these new insights are combined with the
moral hypothesis proposed above—that fairness to future generations requires protection of the integrity of large, self-organizing
systems which provide the context for ongoing human activities—
the result is a neutral management paradigm capable of uniting
environmentalists and resource managers in a broad consensus. It
is neutral in the sense that it avoids literal versions of organicism
and it avoids moral obligations deriving from attributions of
moral value to a supraorganism. The new paradigm, therefore,
avoids both vitalism, which is offensive to many scientists, and
"environmental fascism" (the view that individual interests must
be sacrificed for the good of a reified supraorganism), which is offensive to traditional ethicists who emphasize the moral importance of human individuals (Callicott 1989; Regan 1983).

Those who are reductionistically inclined may also find this
approach acceptable. Reductionistic explanations are important
and can even be claimed to be "complete" in the limited sense that
they provide useful abstract models for understanding the behavior of individual elements that compose a given level of the system—especially insofar as that system is considered in the relatively short frame of time in which individual behavior is manifest. Interactions of individual hunters, wolves, and deer, for example, might be analyzed *annually* by a game-management model
that is mainly reductionistic and aggregative. As trends in the behavior of hunters, the removal of predators, and so forth change
the boundary conditions of the system, the management model
must be expanded to incorporate a larger system, including the
quality of the range and a frame of time measured in decades
(Leopold 1949; Norton 1990).

This hierarchical, multiscalar paradigm, therefore, suggests a
two-stage model for management. *Resource management*, which
applies an essentially reductionistic and economic criterion to individual behavior, applies to the annual cycle of production and
consumption. *Environmental management* which applies a criterion
informed by an ecological understanding of the system that provides annual yields, is governed by constraints necessary to pro-

tect the self-organizing and self-regulatory system that provides the context of annual management decisions. Resource management, therefore, operates within the constraints indicated by an ecological understanding of its context.

This new paradigmatic solution to the long-standing problem of parts and wholes is not without its conceptual "costs," however. Western thinkers as diverse as Moses, Democritus, Plato, Aristotle, Descartes, and Newton have been united by a belief that ultimate explanations cannot be based in the realm of changing and unstable relationships or objects. Dominant Western thinkers have insisted that explanatory sequences end only with eternal and unchanging objects or principles—an eternal God, indestructible atoms, Platonic forms, or fixed essences. But the new ecological paradigm must wean itself from all such sources of stability and security. Processes of nature occur as nonrepeatable, fleeting events within an ever-changing, ever-developing, "open" system. In an open system, value is placed on creativity, self-generation, innovation, adaptability. Stability appears, in the new paradigm, as a (sometimes useful) illusion born of contraposing differing frames of time. Changes in ecological time take place in a "stable" context, but the stability is relative to the larger, dynamic (but still slower-changing,) abiotic environment—the level of the system where dynamics are measured in geological time spans appropriate to plate tectonics and erosion.

While this approach allows us to avoid metaphysical commitments attendant on literal versions of organicism, the implication that the system is changing and developing on multiple levels clearly places it closer to the organicist tradition than to the mechanistic; it embraces a new governing metaphor for environmental discourse and reasoning. But the proposed application of the new metaphor implies no literal reification of one level of the hierarchy as the "organism" of nature, and it may be preferable to choose not the organism but the flame, or some other self-organizing but open system, as the basic metaphor (Earley 1990).

THE PROBLEM OF SCALE

Apart from the heady developments in the theory of self-organizing systems just described, there are also practical reasons for optimism. Schaeffer and Karr have made significant progress in proposing and implementing tests that can serve as early *indicators* of declines in ecosystem health and integrity. (See Schaeffer et al. 1988; Novak and Schaeffer 1990; Karr 1987, 1990, 1991). Applied in the field, these indicators can signal when a system, such as a lake or stream, may be headed for trouble and can trigger more detailed studies of declining integrity.

The problem with these practical indicators is that they have not been related in a clear and general way to the theoretical considerations sketched in Part II of this book. Schaeffer et al. (1988) have noted that many of the useful indicators have been chosen because of their apparent correlation with characteristics of more mature systems, systems that have developed considerably since a major disturbance. While these connections have not been made theoretically precise, there exists a clear possibility of building mathematical and theoretical bridges between current fieldwork and the abstract models of developmental complexity central in the theoretical aspects of the new paradigm. These bridges would exploit the analogies implicit in the developmental view of ecological systems without assuming a single endpoint, or climax state, for open, developing systems in nature.

We should not, however, underestimate the difficulties still remaining. The disadvantage of the minimalist approach described here, as compared to Gaia theory, for example, is that it offers few clues as to which of the infinite number of possible hierarchical arrangements is to be chosen for the purposes of analysis, understanding, and management. That is, the disadvantage of avoiding arbitrary decisions regarding which level of the system is "whole"—and of choosing self-organization, rather than organicism, as the central idea of the new metaphor—is that this minimalist approach leaves the proposed paradigm with no clear guidance regarding the *boundaries* (membranes) of systems for management and no clear physiology of systems to be managed. The concept of a membrane represents, in other words, only a formal category and provides little substantive guidance regarding the proper scale and perspective for characterizing management problems.

There is great interest in analogies to human and veterinary medicine as sources of guidance. At this point, however, it remains unclear how one can move past the helpful analogies in the work of Rapport et al. (1981, 1985) to actual models that represent stability in open systems. The problem is that we lack an adequate theoretical or practical solution to the problem of scale and perspective in environmental management. This problem represents a practical analog of the philosophical conundrum of parts and wholes. Application of indicators of a system's health can only be explained theoretically if we can characterize the physiology of the complex systems of nature, but the physiological characterization of systems awaits resolution of scale issues that have proved recalcitrant. The assumption of a dynamic world has not, as a result of this gap, resulted in easy application of principles in ecological theory to ecological management.

The problem is not a shortage of theories of ecosystem functioning; indeed, we have an embarrassment of riches. Because nature is self-organizing on so many levels, it is easy to state a variety of incommensurable hierarchical arrangements to explain its many aspects, and managers have no ready means to determine which of these many scales to take as basic. Until we have a means to build scale and perspective into our models, there can be no precise explanation of why certain indicators correlate with important ecosystem characteristics. We therefore are hypothesizing that the missing link in the emerging theory and practice of integrated environmental management is an approach to scale that will link theories of ecosystem management with the easily observed indicators of ecosystem health. The effort to complete this linkage is complicated by the foregoing theoretical problems regarding the closed nature of the conventional organic analogies (Norton 1991).

Problems of scale affect environmental management decisions in at least four important ways. The first is *boundary issues*. Every determination of a system's health or degradation presupposes an identification of the system that is of scientific and managerial interest. It is a common complaint that current jurisdictional lines have been drawn with little attention to the natural features of the land or its biogeography. Progress in protecting more holistic management goals, therefore, requires a better understanding of how scientifically adequate ecosystem boundaries can inform and integrate the work of multiple agencies with mandates to manage elements of large ecosystems.

The second is *temporal perspective*. One assumes, upon identifying a whole system for management, that the system maintains an identity through time. But it is also obvious that biological systems are dynamic; they change in many ways and according to multiple frames of time. For example, systems change in geological, climatic, and ecological time; further, each of these scales embodies natural cycles of varying periodicity. It is seldom clear whether managers should be protecting features that are familiar to our culture, ecological features that support certain essential "services," or long-standing features that provide the geological context for ecological processes.

Organizational structure and scale constitute the third set of issues. Even given spatial and temporal boundaries as designations of whole-system units of management, there would remain important questions regarding the substructure of the systems we bound and describe as whole. For example, the decision to manage a bay system encompassing an entire watershed as if it were an organism does not determine which parts should be thought of as vital "organs." Should a bay system be modeled as composed of tributaries and bay proper, or into tidal and nontidal areas, or some other arrangement? Serious attempts to manage a whole system require some designation of organs, which require membranes, however permeable, that represent boundaries at a smaller scale within the system designated as whole.

The fourth problem concerns *scale and degree of impact*. The biotic systems we value are dynamic, but they also maintain equilibria with respect to a number of features essential to those systems we enjoy and depend upon. This relative constancy is essential to allow planning and purposeful action and, on a larger scale, to allow behavioral, cultural, and genetic adaptation. Even human activities that tend to degrade natural systems are damped out and equilibrated, provided that the impacts of these activities are on a sufficiently limited scale that does not exceed the system's assimilative capacity. Decisions of whether or not a certain human activity entails an "acceptable" or "unacceptable" risk of degrading a system, therefore, depend on the scale of these activities in relation to the scale of the system.

The centrality and intransigence of scaling problems are therefore established. While there exists a coherent direction for a theory of contextual management, as well as practical tests that can indicate oncoming "illnesses" in ecological communities, the application of these advances within a coherent approach to management awaits an adequate solution to the problem of managerial scale.

SCIENCE OR POLICY?

It is commonly assumed, apparently, that questions of scale can be resolved "scientifically." Ecological science, on this assumption, informs whole-ecosystem management and, while the problems of scale are considered recalcitrant and messy, it is assumed that progress in ecological science will eventually resolve them.

It may prove necessary to question this assumption. Therefore we ask: what kinds of evidence, scientific or otherwise, are necessary to justify a given choice regarding the scale on which environmental problems are addressed? This is not a question in any particular science; it requires a transdisciplinary, "philosophical" perspective. This point can be illustrated by briefly examining three possible answers to this question.

First is the Naive View. Ecologists study organisms as they relate to their environment, and the environment of all species in an area is the complexly organized ecological system of the area.[6] It constitutes a relatively self-sufficient system ecologically (recognizing, of course, that boundaries are somewhat permeable). While this approach is attractive to advocates of whole-ecosystem management, it is not supported by scientific practice because ecological scientists use many scales, vary them for the convenience of particular studies, and generally treat the ecosystems they study as temporary conveniences to be discarded and replaced at will. The defender of the Naive View, therefore, must explain how the manager is to choose the "correct" scale for management from the myriad possibilities. Even to start such an explanation seems to commit the advocate to a kind of philosophical realism: there exists a single system in nature that can be distinguished from its competitors by observation of the characteristics of real ecological systems. So the Naive View, however attractive, at least needs explanation, clarification, and general shoring up.[7]

The second perspective is Constructivism. If we wish to avoid naive realism—the view that the models and units proposed by ecologists are given in nature—we might instead take the view that ecological systems are models constructed by ecologists in the processes of their studies but, nevertheless, can be sorted according to their usefulness for some overarching scientific purpose, such as "understanding ecosystem functioning" (Allen and Starr 1982). The uniquely correct scale for managers to use, one might then argue, is the scale that allows understanding of the system—

for clearly one cannot manage something one cannot understand. But again, any claims to uniqueness seem to rest on the very doubtful premise that there exists only one correct way to understand a complex ecological system.

The third strategy is Experimentalism. A more promising approach might be to recognize that decisions of scale are really public policy decisions and that, while scientific information is important in resolving them, they are underdetermined by scientific information itself. That is, questions of scale will only be understood if we accept that they involve value judgments. Our management scale is at least partly a function of the goals we adopt in management; but goals cannot, in turn, be understood prior to scientific understanding. Certainly knowledge is a factor in forming ecological attitudes. This is not, however, an insoluble chicken-and-egg problem. In a dynamic system that places humans and their processes of valuing within nature, values will be judged according to adaptability. The task of studying natural processes by experimenting with conservation practices creates a spiral of scientific progress and increasing public knowledge of the importance of ecosystem protection. According to this approach, the apparently limitless number of possible hierarchical models of natural systems will be sorted according to their usefulness in clarifying, explaining, and achieving public goals. The correct scale on which to address a management problem is determined by what society wants to accomplish with that system. The initial goals—and the means to achieve them—are then tested in carefully controlled experiments in system management.

In the Chesapeake Bay area, for example, the consensually accepted goal of improving water quality shapes the decision to treat the entire watershed as the whole system (Horton 1987). Similarly, the policy decision of the National Park Service to manage Yellowstone to maintain viable populations of large mammals implies that they must advocate a whole "bioregion" as the management unit. Two key points arise, however: First, these management initiatives must be regarded as social experiments—experiments in which ecologists play key roles, but not as "pure" scientists who ignore public goals and values. To achieve success, managers, scientists, and the public must work closely together. Second, the Axiom of Differential Fragility implies that the process must be local. Problems of goals and problems of scale cannot be resolved in the abstract; they must emerge from a local process of dialogue and debate. Failures in management, such as disasters in Yellowstone and other national parks, result when managers fail to heed inputs from independent scientists (Newmark 1985) and when management plans are imposed bureaucratically rather than developed in a cooperative and experimental spirit with local interest groups.

Thus an important function of environmental management is to work with local regions to develop, by discussion and careful experimentation, a set of goals that express the local culture even as they protect the contextual features important to its continuance. Conservation biologists, ecosystem restorationists, and ecologists must contribute to this admittedly messy process of social and environmental experimentation. But this approach, however promising, apparently entails the abandonment of claims that conservation biology and ecological economics are value-free sciences—an implication that may cause significant discomfort among professionals (Murphy 1990).

CONCLUSION

A new paradigm of environmental management has been sketched. This paradigm is characterized by an emphasis on the interrelatedness, hierarchical complexity, dynamism, openness, and creativity of systems to be managed. The new paradigm implies that, in management, innovation and adaptation are preferable to fixed principles and that experimentation is essential. Furthermore, the Axiom of Differential Fragility, which recognizes the particularities of systems, entails that ecological study and ecological management must, like medicine, be particularistic. Abstract reasoning and general principles, while sometimes helpful in establishing what to look for in particular situations, are no substitute for careful study of local conditions.

Because ecosystems are self-organizing and creative—but not self-organizing according to a prior plan of either humans or some supraorganism—management must have as a central goal the protection of the system's creativity. Human growth and development are parasitic upon these creative forces, even as management efforts, for good or ill, shape that creativity. This promising paradigm can only be instituted as a general guide to management if conservation biologists, restorationists, and environmental managers reduce their isolation and participate in a public dialogue. Likewise, philosophers, anthropologists, humanists, economists, and citizens must join the search for appropriate public values. In short, all parties must enter the dialogue, rather than creating closed disciplines insulated by jargon and esoteric mathematics. Environmental management, in this view, is basically a policy discipline. Conservation biology, social sciences, and the humanities must therefore interact within the maelstrom of public debate regarding conservation goals, practices, and standards.

References

Allen, T.F.H., and T. B. Starr. 1982. *Hierarchy Theory*. Chicago: University of Chicago Press.

Botkin, D. 1990. *Discordant Harmonies*. New York: Oxford University Press.

Boulding, K. 1991. "What Do We Want to Sustain? Environmentalism and Human Evaluations." In R. Costanza (ed.), *Ecological Economics: The Science and Management of Sustainability*. New York: Columbia University Press.

Callicott, J. B. 1989. *In Defense of the Land Ethic*. Albany: State University of New York Press.

Chase, A. 1986. *Playing God in Yellowstone*. Boston: Atlantic Monthly Press.

Daly, H. E., and J. B. Cobb, Jr. 1989. *For the Common Good*. Boston: Beacon Press.

Earley, J. 1990. "The Rock and the Flame." Paper presented at the American Philosophical Association meetings, Boston.

Horton, T. 1987. "Remapping the Chesapeake." *The New American Land* 7: 7–16.

Joseph, L. E. 1990. *Gaia*. New York: St. Martin's Press.

Karr, J. R. 1987. "Biological Monitoring and Environmental Assessment: A Conceptual Framework." *Environmental Management* 11:249–256.

———. 1990. " Biological Integrity and the Goal of Environmental Legislation: Lessons for Conservation Biology."*Conservation Biology* 4:244–250.

———. 1991. "Biological Integrity: A Long Neglected Aspect of Water Resource Management." *Ecological Applications* 1(1):66–84.

Koestler, A. 1967. *The Ghost in the Machine*. New York: Macmillan.

Kuhn, T. 1970. *The Structure of Scientific Revolutions*. Chicago: University of Chicago Press.

Leopold, A. 1939. "A Biotic View of Land." *Journal of Forestry* 37:727–730.

———. 1949. *A Sand County Almanac*. New York: Oxford University Press.

Lovelock, J. E. 1977. *Gaia: A New Look at Life on Earth*. New York: Oxford University Press.

———. 1988. *The Ages of Gaia*. New York: Norton.

Murphy, D. D. 1990. "Conservation Biology and Scientific Method." *Conservation Biology* 4:203–204.

Meine, C. 1988. *Aldo Leopold: His Life and Work*. Madison: University of Wisconsin Press.

Newmark, W. D. 1985. "Legal and Biotic Boundaries of Western North American National Parks: A Problem of Congruence." *Biological Conservation* 33:197–208.

Norton, B. G. 1977. *Linguistic Frameworks and Ontology*. The Hague: Mouton.

———. 1987. *Why Preserve Natural Variety?* Princeton: Princeton University Press.

———. 1989. "Intergenerational Equity and Environmental Decisions: A Model Using Rawls' Veil of Ignorance." *Ecological Economics* 1:137–159.

———. 1990. "Context and Hierarchy in Aldo Leopold's Theory of Environmental Management." *Ecological Economics* 2:119–127.

———. 1991a. "Ecosystem Health and Sustainable Resource Management." In R. Costanza (ed.), *Ecological Economics: The Science and Management of Sustainability*. New York: Columbia University Press.

———. 1991b. *Toward Unity Among Environmentalists*. New York: Oxford University Press.

Novak, E. W., and D. J. Schaeffer. 1990. "Developing Comprehensive Field Studies to Identify Subchronic and Chronic Effects of Chemicals on Terrestrial Ecosystems: Ecosystem Health-VI." In S. S. Sandhu et al., *In Situ Evaluations of Biological Hazards of Environmental Pollutants*. New York: Plenum Press.

O'Neill, R. V., D. L. DeAngelis, J. B. Waide, and T.F.H. Allen. 1986. *A Hierarchical Concept of Ecosystems*. Princeton: Princeton University Press.

Page, T. 1991. "Sustainability and the Problem of Valuation." In R. Costanza (ed.), *Ecoloical Economics: The Science and Management of Sustainability* New York: Columbia University Press.

Prigogene, I., and I. Stengers. 1984. *Order Out of Chaos*. New York: Bantam Books.

Rapport, D. J. 1984. "State of Ecosystem Medicine." In V. W. Cairns, P. V. Hodson, and J. O. Nriagu. *Contaminant Effects on Fisheries*. New York: Wiley.

———. 1989. "What Constitutes Ecosystem Health?" *Perspectives in Biology and Medicine* 33(1):120–132.

Rapport, D. J., H. A. Regier, and C. Thorpe. 1981. "Diagnosis, Prognosis, and Treatment of Ecosystems under Stress." In G. W. Barrett and R. Rosenberg, *Stress Effects on Natural Ecosystems*. New York: Wiley.

Rapport, D. J., H. A. Regier, and T. C. Hutchinson. 1985. "Ecosystem Behavior Under Stress." *The American Naturalist* 125:617–640.

Rawls, J. 1971. *A Theory of Justice*. Cambridge: Belknap Press.

Rees, W. R. 1990. "The Ecology of Sustainable Development." *The Ecologist* 20:18–23.

Regan, T. 1983. *The Case for Animal Rights*. Berkeley: University of California Press.

Sagoff, M. 1988. *The Economy of the Earth*. Albany: State University of New York Press.

Schaeffer, D. J., E. E. Herricks, and H. W. Kerster. 1988. "Ecosystem Health. I: Measuring Ecosystem Health." *Environmental Management* 12:445–455.

Toulmin, S. 1961. *Foresight and Understanding*. New York: Harper Torchbooks.

———. 1972. *Human Understanding* Vol. 1. Princeton: Princeton University Press.

Ulanowicz, R. E. 1986. *Growth and Development*. New York: Springer-Verlag.

Notes

1. In economics, see Daly and Cobb (1989) and Page (1991); in ecology, see Botkin (1990); in philosophy, see Sagoff (1988) and Norton(1991).
2. The idea that mature science operates against a backdrop of shared and unquestioned assumptions—a paradigm—was popularized by Kuhn (1962, 1970). Kuhn's original formulations sometimes suggested that paradigmatic change is not susceptible to rational analysis, although Kuhn has since distanced himself from antirationalism. Here we use the concept in the less radical form defended by Toulmin (1961).
3. It is preferable to avoid the common term "autonomous," because its root "nomos" suggests self-government or self-legislation, which may be taken to have intentional implications. A better term to express

neutral self-organization is "autopoiesis," which refers to self-creation and thereby lacks the implication of intentional self-government. See Rees (1990).

4. Total diversity (gamma diversity) is defined as a function of within-habitat diversity (alpha diversity) and cross-habitat diversity (beta diversity). It is the diversity characteristic of a landscape composed of many habitats and microhabitats. This is not, of course, to deny that species composition does change through time; it might, therefore, be preferable to think not of species compositions but of processes of competition and substitution that are ecologically driven—a system's trajectory of change, that is, mainly originates from within the system and changes in ecological time. See Norton (1987).

5. It is important to note that Lovelock explicitly recognized that the biosphere is an open system. Nevertheless, he personalizes the atmosphere as the hierarchy that exerts agency.

6. "Naive" is not meant here as a pejorative term—sometimes the naive answer is the best answer.

7. See Chase (1986) for an argument against the Naive View; he uses it as an argument, however, for abandoning eosystem management altogether.

2

Aldo Leopold's Metaphor

J. Baird Callicott

The concept of ecosystem health is metaphorical. *Health* in the literal sense of the term characterizes only a state or conditon of an organism. Plato, however, provides a notable historical precedent for using the concept of health metaphorically. In *The Republic*, Socrates says, "Then virtue is the health and beauty and well-being of the soul, and vice the disease and weakness and deformity of the same" (Plato 1937:709). Plato also provides an instructive historical precedent for why one would want to use the health metaphor. Socrates next asks whether or not a person would prefer to live in accordance with virtue or its opposite—the fundamental question of the whole dialogue. And Glaucon replies: "In my judgment, Socrates, the question now becomes ridiculous. We know that, when the bodily constitution is gone, life is no longer endurable; and shall we be told that when the very essence of the vital principle is undermined and corrupted, life is still worth having to a man? (Plato 1937:709).

The concept of health, in both its literal and figurative senses, is at once descriptive and prescriptive, objective and normative. Health, literally, is an objective condition of an organism capable of more or less precise empirical description. But it is also an intrinsically valuable state of being in the sense of C. I. Lewis (1946).[1] Except under the most unusual circumstances, it is never better to be sick than well. Indeed, for Plato, health is the "good" of the body, its appropriate condition of internal order or organization. Similarly, Plato argues, virtue is the good of the soul, its appropriate condition of internal order or organization. And, further extending the same metaphor, Plato also argues that justice is the intrinsically good, healthy state of the body politic. Today, we could add the ecosystem to Plato's series of analogies. Ecosystem health is a condition of internal order and organization in ecosystems, which no less than analogous conditions of body, soul, and society are both intrinsically good and objective (and specifiable in principle).

Echoing Plato, Peter Miller (1981:194) explains why ecosystem health is a good candidate for intrinsic valuation: "It wears the appearance of (a) a value, a basic ingredient in living well or thriving, which (b) characterizes non-human and even non-conscious living [systems] (c) independently of human utilities." Thus, as we see, if the health metaphor may be plausibly and persuasively extended to ecosystems, then the fact/value (is/ought) dichotomy that has routinely plagued policy debates and applied science can be obviated.

Here, of course, we must be cautious. Assuming the distinction between the subjective and objective realms—common to both modern philosophy and modern classical science—all value depends on a subjective valuer. But as Holmes Rolston III (1981:114) has eloquently pointed out, what we value is not at all capricious: "We make something a target merely by aiming at it. But our interest in apples is not so arbitrary. It depends in part on something that is found there." Even our purely instrumental valuing depends on fact, or on what Rolston calls "descriptions." We human beings are also capable of noninstrumental valuing. Bodily health is a paradigm case of something that is intrinsically as well as instrumentally valuable; it is good in and of itself, as well as a necessary condition for getting on with our projects. So is virtue and a healthy body politic, Plato would argue. And had the concept of an ecosystem been current in Plato's day, doubtless he would have also argued that ecosystem health too is something valuable for its own sake as well as for its utility (see Chapter 3 of this volume). We have a reasonably clear and detailed scientific description of bodily health. A new and socially important task for science is to tackle the problem of specifying objective norms of ecosystem health. Like the medical norms of bodily health, the norms of ecosystem health would be simultaneously descriptive and prescriptive, objective and nomothetic, instrumentally and intrinsically valuable.

If the concept of ecosystem health turns out to be plausible and persuasive and if its norms and indices can be specified, the cause of biological conservation may be bolstered. Presently it is being undermined by an insidious skepticism within the scientific community. The erstwhile benchmarks of biological conservation are under withering attack.

THE CRUMBLING CORNERSTONES OF BIOLOGICAL CONSERVATION

Formerly, one could argue for conserving biodiversity because it was believed to be causally related to ecological stability. The "diversity-stability hypothesis," however, has been severely criticized (May 1973; Goodman 1975)—how decisively, I am not competent to judge. But the mandate for the conservation of biodiversity will be put at risk to the extent that one of its principal rationales is suspect.

Wilderness once served as a gross standard of conservation. Wilderness was imagined to be a pristine environment, defined by an act of the United States Congress as "in contrast to those areas where man and his own works dominate the landscape, an area where the earth and its community of life are untrammeled by man, where man himself is a visitor who does not remain" (Nash 1967:5). But wilderness preservation, so understood, assumes a pre-Darwinian religious and metaphysical separation of *Homo sapiens* from the rest of nature. After Darwin, we cannot suppose that "man" is anything but a precocious primate, a denizen of "the earth" and a member of its "community of life." "His own works," therefore, are as natural as those of termites or beavers. Biological conservation via wilderness preservation is also vitiated by ethnocentrism. To suggest that, prior to being "discovered" only half a millennium ago by the European subspecies of *Homo sapiens*, any landscape in North or South America was "untrammeled by man, where man himself is a visitor," implies either that large portions of North and South America were uninhabited or that the aboriginal inhabitants of the New World had no significant intentional or even unintentional effect upon their lands. Or, worse, it could imply that American Indians were not truly human, were not "man."

But by 1492, all of North and South America were fully if not densely populated (Dobyns 1966). And the effects of more than 10,000 years of human inhabitation of the Western Hemisphere have been profound and, on the eve of European encroachment, were ongoing (Cronon 1983). After the arrival of *Homo sapiens* in the Western Hemisphere, some ten millennia ago, the only large land area fitting Congress' description of wilderness was Antarctica (and now a good bit of that continent, and the atmosphere above it, have been thoroughly trammeled). When an area's aboriginal human inhabitants are removed, in order to create a wilderness, the ecological conditions that existed at the time of their removal, presumably the conditions to be preserved in their "virgin" state, are put at jeopardy (Bonnicksen 1990). Biological conservation via wilderness preservation thus proves to be based on an incoherent idea: the wilderness idea.

Further, whether a "natural" area has been dominated by *Homo sapiens* or not, in time it will nevertheless change. The fourth dimension of ecosystems has recently been emphasized by Daniel Botkin (1990). Nature is dynamic. Change at every frequency— diurnal, meteorological, seasonal, successional, climatic, evolutionary, geological, astronomical—is inevitable. According to Botkin, the concept of succession in ecology culminating in a climax community that will perpetuate itself generation after generation until reset by wind, fire, chain saw, plow, or some other disturbance is suspect. While accounting for change, the concept of succession-to-climax, he argues, posits, like Aristotle's theory of motion, a static state as the "natural" condition of ecosystems.

Presenting a more specific and subtle challenge to the rationale for biological conservation, change in the structure of biotic communities has been recently emphasized. According to Michael Soulé (1990:234),

> shifts in scientific fashion will facilitate the transition between the traditional view of biogeographic integrity and the postmodern acceptance of cosmopolitanization. . . . The acceptance of the "individualistic paradigm" of community composition . . . which posits that collections of species that exist in a particular place are a matter of historical accident and species-specific, autecological requirements . . . is reinforced by analyses of Holocene distributions

of contemporary species. These studies are undermining typological concepts of community composition, structure, dynamics, and organization by showing that existing species once constituted quite different groupings or "communities."

These recent developments—the impeachment of the diversity-stability hypothesis, the evaporation of the wilderness idea, the diminishing credibility of the Clementsian holistic paradigm and the corresponding ascendency of the Gleasonian individualistic paradigm in theoretical ecology, the impeachment of the community succession-to-climax model and even the typological community, and the emphasis, in general, on change rather than continuity (a kind of neocatastrophism, as it were, supplanting uniformitarianism)—all give aid and comfort to the foes of biological conservation. If change is a fundamental feature of nature; if humans are a part of nature and anthropogenic changes are as natural as any other; if, for more than 10,000 years, there have been no large-scale, pristine, untouched terrestrial wilderness environments (outside Antarctica); if species in communities can mix and match as they always have to form novel associations; if diversity is not necessarily essential to stability; then how can anyone express more than a personal subjective preference in declaring any change whatever that human beings may impose on landscapes to be bad? What is wrong, objectively wrong, with urban sprawl, oil slicks, global warming, or, for that matter, abrupt, massive, anthropogenic species extinction—other than that these things offend the quaint tastes of a few natural antiquarians? Most people prefer shopping malls and dog tracks to wetlands and old-growth forests. Why should their tastes, however vulgar, not prevail in a free market and democratic polity? Kristin Shrader-Frechette (1989:76) has explicitly brought us to this omega point:

> Ecosystems regularly change and regularly eliminate species. How would one . . . argue that humans should not modify ecosystems or even wipe out species, for example, when nature does this itself through natural disasters, such as volcanic eruptions and climate changes like those that destroyed the dinosaurs? . . . One cannot obviously claim that it is wrong on ecological grounds for humans to do what nature does—wipe out species.

The concept of ecosystem health to the rescue! Yes, change is natural, human beings are a part of nature, and anthropogenic changes are no different from other natural changes. But, quite irrespective of the vagaries of taste, we may still argue that some are bad and others good, if we can specify objective norms of ecological health against which we may evaluate human modifications of the landscape. By the same criteria, of course, we might evaluate the changes wrought by any other species.

LEOPOLD'S CONCEPT OF LAND HEALTH

I leave to ecologists the task of determining whether or not the concept of ecosystem health makes scientific sense and, if it does, what its general characteristics and indices might be. As Plato provides a *locus classicus* for turning the metaphorical concepts of psychological and political health to moral advantage, Aldo Leopold provides a *locus classicus* for turning the metaphorical concept of ecosystem health (or "land health," as he more simply denominated it) to conservation advantage. That so distinguished and prescient a conservationist as Leopold thought that the concept had promise, and envisioned its scientific articulation, may be taken as prima facie evidence that the notion is at least worth a close look.

During the last decade of his life, 1938–1948, Leopold frequently employed the concept of land health in his sundry writings, including several of the essays in *A Sand County Almanac*. In two of his papers from this period, Leopold provides a sustained discussion of the concept of land health. Ironically, in view of the foregoing remarks, the first is "Wilderness as a Land Laboratory," in which Leopold offers a novel argument for wilderness preservation. Here he suggests that wilderness may serve as "a base-datum for land health" and defines "land health" as nature's capacity for "self-renewal," a definition that he reiterates in subsequent usages and a definition that carries, importantly, both dynamic and functional, rather than static and structural, connotations (Leopold 1941:3). Though not invoking Clements' organismic ecological paradigm in any strict or specific sense, Leopold here, as elsewhere, closely associates the concept of land health with an organic image of nature:

There are two organisms in which the unconscious automatic processes of self-renewal have been supplemented by conscious interference and control. One of these is man himself (medicine and
public health) and the other is land (agriculture and conservation).
[Leopold 1941:3].

That Leopold introduces the concept of land health, not as a ca-
sual rhetorical device, but as a serious scientific project, is suggested by the way he explores the analogy that he here draws to
medicine. In the field of medicine, the symptoms of disease are
manifest and doctoring is an ancient art, but medical science is
relatively young and still incomplete. Analogously, "the art of
land doctoring is being practiced with vigor," he comments, "but
the science of land health is a job for the future" (1941:3). In 1941,
ecology was not capable of specifying the norms of land health.
On the other hand, the symptoms of "land sickness" were all too
evident to the discerning conservationist. Among such symptoms,
Leopold mentions soil erosion and loss of fertility, hydrological
abnormalities, and the occasional irruptions of certain species and
the mysterious local extinctions of others.

While he claims that the most perfect "base-datum of normal-
ity" is wilderness, Leopold does not argue that the only way for
land to stay healthy is to remain in an untrammeled condition.
One may find places "where land physiology remains largely
normal despite centuries of human occupation" (1941:3). He believed the well-watered regions of Europe to be such places.
Indeed, the practical raison d'être for a science of land health is
precisely to determine the ecological parameters within which
land may be humanly occupied without making it dysfunctional,
just as the whole point—or at least the only point that Leopold
makes in this paper—of wilderness preservation is to provide a
land laboratory in which such a science might be explored.

Leopold's uncritical belief in the existence of wilderness is
traceable, I surmise, to the unconscious ethnocentrism that he
shared with most of his contemporaries. In a paper written during
the same year, he remarks that "the characteristic number of
Indians in virgin America was small" (Leopold 1991a:282). And in
a paper published four years earlier he rhetorically reduces one
group of American Indians, "the predatory Apache," to a form of
less than fully human indigenous wildlife (Leopold 1937:118). But
of course North America, no less than temperate Europe, had also

been subjected to "centuries of human occupation," indeed, more than 100 centuries.

Leopold's other sustained discussion of "land health" is found in an (until now) unpublished 1944 report, "Conservation: In Whole or in Part?" In it he defines conservation as "a state of health in the land" and land health, once again, "as a state of vigorous self-renewal" (Leopold 1991b:310). Here Leopold expressly draws out the functional connotation of this definition: "Such collective functioning of interdependent parts for the maintenance of the whole is characteristic of an organism. In this sense land is an organism, and conservation deals with its functional integrity, or health" (p. 310). The maintenance of land health therefore is not necessarily the same thing as maintenance of existing community structures with their historical complement of species. Exotics may immigrate on their own or be deliberately introduced (cautiously) and evaluated, not xenophobically, but on the basis of their impact on the functional integrity of the host community. They may be pathological, they may be benign, or, conceivably, they may actually enhance ecosystem functions.

In "Conservation: In Whole or in Part?" Leopold affirms the importance of diversity for ecological function. Referring to the postglacial upper Midwest he writes: "The net trend of the original community was thus toward more and more diversity of native forms, and more and more complex relations between them" (1991b:312). He then draws the classic, but presently impugned, connection between diversity and stability: "Stability or health was associated with, and perhaps caused by, this diversity and complexity" (p. 312). It is a tribute to Leopold's scientific sensibilities that he carefully avoids stating dogmatically that stability is caused by diversity. Indeed, he registers an express caveat: "To assert a causal relation would imply that we understand the mechanism." But, absent thorough understanding, he argues (p. 315):

> The circumstantial evidence is that stability and diversity in the native community were associated for 20,000 years, and presumably depended on each other. Both now are partly lost, presumably because the original community has been partly lost and greatly altered. Presumably, the greater the losses and alterations, the greater the risks of impairments and disorganizations.

As the science of land health is, for Leopold in the 1940s, only envisioned, only programmatic, he suggests that the art of land doctoring can only proceed on such circumstantial evidence and err on the side of caution. The rule of thumb for ecological conservation then should be, he thinks, that "the land should retain as much of its original membership as is compatible with human land-use [and] should be modified as gently and as little as possible" (p. 315). But, again, it does not take a well-developed science of land health to notice the symptoms of land illness. Apart from those already mentioned in his earlier paper, Leopold here adds the qualitative deterioration in farm and forest products, the outbreak of pests and disease epidemics, and boom and bust wildlife population cycles. Further to a governing philosophy of ecological conservation, Leopold suggests something similar to what is known today as holistic and preventive, as opposed to reductive and invasive, medicine: "This difference between gentle and restrained, as compared with violent and unrestrained, modification of the land is the difference between organic and mustard-plaster therapeutics in the field of land-health" (p. 315). Leopold then goes on to outline a unified and holistic conservation strategy, as the title of his paper would indicate.

LAND HEALTH AND AUTOPOIESIS

The contemporary ecologist looking for substantive norms of ecosystem health that might serve as objective criteria for the evaluation of human modifications of historic ecological conditions will find a review of Leopold's remarks about land health unrewarding. Leopold's scientific scruples preempted any impulse he may have had speculatively to detail them in the absence of basic ecological research. In general, he closely associates land health with both integrity—which he seems to understand primarily structurally—and stability. Leopold equates the integrity of land with the continuity of stable communities over long periods of time. Such integrity and stability, he cautiously suggests, depend on species diversity and the complexity of relations between native species. In short, we find in Leopold only today's conventional environmental wisdom, which then was new and fresh but is now a bit tired and tarnished. On the other hand, because Leopold's general definition of land health is more functional than structural, more dynamic than static, it is very much in tune with current fin de siècle ecological theory. And it is also very much in tune with the growing emphasis—evident in later chapters of this book—on complexity rather than simple diversity measures, which may point a more promising direction.

In fact, there exists a notion in contemporary biology cognate with Leopold's idea of land health: the concept of autopoiesis. Leopold's concept of land health and the concept of autopoiesis may indeed be mutually reinforcing: the imprimatur of Leopold's enormous reputation could attract greater attention to, and foster greater exploration of, the concept of autopoiesis; the concept of autopoiesis could update and articulate what Leopold was struggling to express with his notion of land health characterized as the "capacity of land for self-renewal."

Autopoiesis is transliterated from two combined Greek words meaning "self-making." It was coined in 1972 by the Chilean biologists Humberto Maturana and Francisco Varela (1980) to characterize living systems more inclusive than organisms. Thus an organism is autopoietic, but so too are other biological entities. Instead of assimilating supraorganic biological systems to organisms, and implausibly attributing to them all sorts of similar characteristics, à la Clements, the concept of autopoiesis permits a more limited comparison between organisms proper and larger living systems.

Explicitly assimilating Leopold's informal notion of land health with the contemporary formal concept of autopoiesis could, therefore, rescue the former from guilt by association with the discredited organismic ecological paradigm. Though Leopold, for rhetorical effect, will sometimes say that "land is an organism," just as Clements said that a plant association is an organism, in trying to articulate what he means by "land health" Leopold focuses on the capacity of land for self-renewal and more carefully says that "in this sense land is an organism." Ecosystems and organisms, in other words, are very different and—here it seems that Leopold may have been aware—one would be mistaken to think that the former are just larger and more diffuse versions of the latter. Rather, Leopold thinks, they have one very fundamental, not to say essential, characteristic in common: the capacity for self-renewal. More technically expressed, organisms and ecosystems are both autopoietic: self-organizing and self-recreating.

The concept of autopoiesis wears its dynamism on its sleeve, since *poiesis* is a verb. *Health,* on the other hand, is a noun and may therefore suggest a static condition in both organisms and ecosystems. But health, despite the grammar of its name, actually is very much a process, a process of self-maintenance and self-regeneration. In a healthy multicellular organism there is a constant turnover of cells. Further, while there is continuity in the organization of transient cells, over a lifetime there is dramatic ontogeny: from zygote to embryo to fetus to infant; then growth, maturation, reproduction; and finally decline and senescence. Or even more dramatically: from ovum to larva to pupa to imago. Today, ecologists emphasize that ecosystems also change over time, but, like healthy organisms, healthy ecosystems maintain a certain continuity and order in the midst of change. Thus radical and discontinuous change is as destructive of autopoietic ecosystems as it is damaging to autopoietic organisms.

All the symptoms of land illness that Leopold notes are failures of ecological function, not merely alterations in structure. While Leopold believed that the structural integrity of biotic communities guarantees their continued ecological function, he acknowledged that functional continuity could be maintained in the midst of gradual and orderly structural change. According to Maturana and Varela, an autopoietic system, though by definition a system whose whole and sole business is to regenerate itself, ordinarily undergoes structural changes. But those changes must be orderly and, especially if imposed from outside the system (what they call "deformations"), limited: "Any structural changes that a living system may undergo maintaining its identity must take place in a manner determined by and subordinate to its defining autopoiesis; hence in a living system loss of autopoiesis is disintegration as a unity and loss of identity, that is death" (Maturana and Varela 1980:112).

Hence human inhabitation and transformation of ecosystems are not, in principle, incompatible with their health, their autopoiesis, or capacity for self-renewal. Objectively good anthropogenic change is change that benefits people and maintains land health. Objectively bad anthropogenic change is change that results in land sickness or worse in the death of ecosystems. The overgrazed, eroded, and unrestorable erstwhile grasslands of the American Southwest are very sick. The barren, laterized soil of an erstwhile tropical rain forest is dead.

Leopold accepted the wilderness myth wholeheartedly. Indeed, he was one of the most outspoken advocates of wilderness preservation and one of the architects of the North American wilderness movement. Nevertheless, unlike a Muir or a Murie, he devoted himself primarily to the conservation of humanly occupied and used ecosystems—"a more important and complex task," as he put it (1991c:227). And Leopold certainly acknowledged that long and densely populated and heavily used land could be healthy. In "A Biotic View of Land," he writes:

> Western Europe, for example, carries a far different pyramid than Caesar found there. Some large animals are lost; many new plants and animals are introduced, some of which escape as pests; the remaining natives are greatly changed in distribution and abundance. Yet the soil is still fertile, the waters flow normally, the new structure seems to function and persist [1939:729].

Susan L. Flader (1974) observes that during the 1930s Leopold underwent a fairly sudden and dramatic shift of attitude toward land management. From the reductive, Cartesian method, set out in *Game Management*, of identifying and manipulating "factors" affecting wildlife populations—such as food, cover, and predation—Leopold shifted to a more holistic, organic approach. Eugene C. Hargrove (1989, this volume) compares his new attitude to "therapeutic nihilism" in medicine. This nineteenth-century school of medicine frankly acknowledged that doctors then did not know enough about the physiology of the human organism to be confident that any medical manipulation, invasion, or prescription would do more good than harm. So, they argued, doctors should err on the side of caution, do nothing, and hope their patients would recover on their own. Leopold, similarly, believed that the contemporaneous state of ecological knowledge was so incomplete that any humanly imposed changes on land were altogether unpredictable. Hence, he counseled caution and argued that the continued functioning of ecosystems could best be assured by preserving and, where practicable, restoring their historic structural integrity.

CONCLUSION

I have no idea how much more confident we can be today in the state of ecological knowledge. I can say with assurance, however, that neither manipulative management nor economic development are incompatible, in principle, with Leopold's general concept of land health. Human economic activities and the presence of domestic or exotic species are not inconsistent with land health as Leopold conceived it—provided that they do not disrupt ecosystem functions, as indicated by the presence or absence of the symptoms of land-illness (some of which Leopold enumerates).

Of course, Leopold's land ethic may provide other reasons for exercising caution and more stringent constraints on human economic activities. If ecology, for example, could assure us that replacing golden trout with brown in California waters would have no adverse impact on the health of affected aquatic ecosystems, the "biotic right" of species to continuation, advocated by Leopold in "The Land Ethic," might still constitute a compelling reason not to do so. One can imagine all sorts of unnecessary and disfiguring operations that an unscrupulous doctor might perform on an unwitting patient, none of which ultimately compromised the patient's health, to satisfy the doctor's own whims or economic interests. But such operations would certainly compromise the patient's dignity and would violate the patient's rights. Human economic activities should be constrained by considerations of land health. The observance of such constraints is as much a matter of morality as of prudence, since health—personal, social, and ecological—is an intrinsic value. But they should also be constrained by additional ethical and aesthetic considerations—by the biotic right of species to continuance and by the beauty of the biotic community.

References

Bonnicksen, T. M. 1990. "Restoring Biodiversity in Park and Wilderness Areas: An Assessment of the Yellowstone Wildfires." In A. Rassmusen (ed.), *Wilderness Areas: Their Impacts*. Logan: Utah State University Cooperative Extension Service.

Botkin, D. 1990. *Discordant Harmonies: A New Ecology for the Twenty-first Century*. New York: Oxford University Press.

Callicott, J. B. 1982. "Hume's Is/Ought Dichotomy and the Relation of Ecology to Leopold's Land Ethic." *Environmental Ethics* 4:163–174.

———. 1991. "The Wilderness Idea Revisited: The Sustainable Development Alternative." *Environmental Professional* 13: 235–247.

Cronon, W. 1983. *Changes in the Land*. New York: Hill & Wang.

Dobyns, H. F. 1966. "Estimating Aboriginal American Population: An Appraisal of Techniques with a New Hemispheric Estimate." *Current Anthropology* 7:395–412.

Flader, S. L. 1974. *Thinking Like a Mountain: Aldo Leopold and the Evolution of an Ecological Attitude Toward Deer, Wolves, and Forests*. Columbia: University of Missouri Press.

Goodman, D. 1975. "The Theory of Diversity-Stability Relationships in Ecology."*Quarterly Review of Biology* 50:237–266.

Hargrove, E. C. 1989. *Foundations of Environmental Ethics*. Englewood Cliffs: Prentice-Hall.

Leopold, A. 1937. "Conservationist in Mexico." *American Forests* 43:118–120.

———. 1939. "A Biotic View of Land." *Journal of Forestry* 37:727–730.

———. 1941. "Wilderness as a Land Laboratory." *Living Wilderness* 6 (July):3.

———. 1991a. "Ecology and Politics." In S. L. Flader and J. B. Callicott (eds.), *The River of the Mother of God and Other Essays by Aldo Leopold*. Madison: University of Wisconsin Press.

———. 1991b. "Conservation: In Whole or in Part? In S. L. Flader and J. B. Callicott (eds.), *The River of the Mother of God and Other Essays by Aldo Leopold*. Madison: University of Wisconsin Press.

———. 1991c. "Wilderness." In S. L. Flader and J. B. Callicott (eds.), *The River of the Mother of God and Other Essays by Aldo Leopold*. Madison: University of Wisconsin Press.

Lewis, C. I. 1946. *An Analysis of Knowledge and Valuation*. La Salle, Illinois: Open Court.

Maturana, H. R., and F. J. Varela. 1980. "Autopoiesis: The Organization of the Living." In H. R. Maturana and F. J. Varela, *Autopoiesis and Cognition*. Boston: D. Reidel.

May, R. M. 1973. *Stability and Complexity in Model Ecosystems*. Princeton: Princeton University Press.

Miller, P. 1981. "Is Health an Anthropocentric Value?" *Nature and System* 3:193–207.

Nash, R. 1967. *Wilderness and the American Mind*. New Haven: Yale University Press.

Plato. 1937. *The Republic*. In B. Jowett (trans.), *The Dialogues of Plato*. vol. 1. New York: Random House.

Rolston, Holmes, III. 1981. "Values in Nature." *Environmental Ethics* 3:113–128.

Shrader-Frechette, K. 1989. "Ecological Theories and Ethical Imperatives: Can Ecology Provide a Scientific Justification for the Ethics of Environmental Protection?" In W. R. Shea and B. Sitter (eds.), *Scientists and Their Responsibility*. Canton: Watson.

Soulé, M. E. 1990. "The Onslaught of Alien Species, and Other Challenges in the Coming Decades." *Conservation Biology* 4:233–243.

Notes

1. By "intrinsic value" Lewis did not mean what the phrase usually means in contemporary environmental ethics: the objective value (noun) that an entity is alleged to possess independently of any valuing consciousness. Lewis meant, rather, a condition or state of being that one values (verb) for itself.

3

Has Nature a Good of Its Own?

MARK SAGOFF

Many of us recall from childhood—or from reading to our own children—E. B. White's story of the spider Charlotte and her campaign to save Wilbur, a barnyard pig. Charlotte wove webs above Wilbur's sty proclaiming the pig's virtues in words—"TERRIFIC," "RADIANT," "HUMBLE"—she copied from newspaper advertisements salvaged by a helpful rat. Wilbur won a special prize at the county fair. Moved by this miracle, Zuckerman, the farmer who owned Wilbur, spared him from being sent to market. Charlotte saved Wilbur's life.

"Why did you do all this for me?" the pig asks at the end of *Charlotte's Web*. "I don't deserve it. I've never done anything for you."

"You have been my friend," Charlotte replied. "That in itself is a tremendous thing. I wove my webs for you because I liked you. After all, what's a life, anyway? We're born, we live a little while, we die. A spider's life can't help being something of a mess, what with all this trapping and eating flies. By helping you, perhaps I was trying to lift up my life a little. Heaven knows, anyone's life can stand a little of that."

THE VARIETIES OF GOODNESS

Charlotte's Web explores three ways we value nature. First, nature is useful: it serves a purpose, satisfies a preference, or meets a need. This is the instrumental good. Traders who bid on pork belly futures have this kind of value in mind. We know how much a thing is worth in this sense when we know its price.

Second, we may value nature as an object of knowledge and perception. This is the aesthetic good. The judges who awarded Wilbur a prize recognized in him the noble or admirable qualities that distinguished him from other pigs. While the basis of instrumental value lies in our wants and inclinations, whatever they may be, the basis of aesthetic value lies in the object itself—in qualities that demand an appreciative response from informed and discriminating observers. These qualities, for those knowledgeable enough to discern them, made Wilbur a pig to be appreciated rather than to be consumed.

Third, we may regard an object (as Charlotte did Wilbur) with love or affection. Charlotte's love for Wilbur included feelings of altruism, as we would expect, since anyone who loves a living object (we might include biological systems and communities) will take an interest in its welfare. Objects that exemplify ideals, aspirations, and commitments that "lift up" one's life also may evoke love, for example, by presenting goals that go beyond the pursuit of one's own welfare. We might speak of "love of country" in this context. Objects of our love and affection have a moral good and, if they are living, a good of their own.

Aesthetic value depends on qualities that make an object admirable of its kind; when these qualities change, the aesthetic value of the object may change with them. With love, it is different. Shakespeare wrote that love alters not when it alteration finds, and even if this is not strictly true, love still tolerates better than aesthetic appreciation changes that may occur in its object.

Although love is other-regarding in that it promotes the well-being of its object, it does not require actions to be entirely altruistic. Only saints are completely selfless, and it is hardly obvious that we should try to be like them. Nevertheless, anyone's life can stand some dollop of other-regarding, idealistic, or altruistic behavior, as Charlotte says.

When we regard an object with appreciation or love, we say it has intrinsic value, by which we mean that we value the object itself rather than just the benefits it confers on us. We care about these objects because of what they mean to us and express about us, not because of what they do for us. This essay concerns the intrinsic value of nature in its relation to environmental policy and the role of ecology as a scientific discipline in shaping that policy.

ECOLOGICAL HEALTH AND INSTRUMENTAL VALUE

Early this century, when Wilbur was growing up in rural Maine, pigs lived in three kinds of environments. First, wild boars roamed some forested areas; indeed, boars are a problem today in places like the Smoky Mountain National Forest. Second, farmers raised barnyard pigs for their own and their neighbors' consumption. For the most part, however, pigs were mass produced in feedlots concentrated in the Midwest. By 1870, the stockyards of Chicago, for example, processed over one million pigs a year.

It seems to me that concepts of ecological health and integrity— including notions of diversity, resilience, hierarchy, complexity, and so on—may apply to forests in which wild boars roam and perhaps even to Zuckerman's farm but not to corporate feedlots and slaughterhouses. These latter seem as distanced from nature as anything can be; they have more in common with enterprise than with evolution. Yet it is plain that when it comes to putting meat on the table, we rely less on the methods described by Charles Darwin than on those pioneered by Philip Armour and Gustavus F. Swift.

The question I wish to consider is why we should value the health or the integrity of ecological communities and systems. Why should we care about wild forests, for example, as distinct from faster-growing biotechnology-based silvicultural plantations? Why should we care about Mr. Zuckerman's farm rather than about the condominiums and vacation homes that developers might raise on that acreage? Do we value the health and integrity of ecological communities for instrumental reasons? Or do these concepts correlate instead with the qualities of natural systems we find intrinsically valuable?

The health and integrity of ecological communities correlate well—for logical as well as empirical reasons—with the intrinsic value of those communities. Biological integrity is not a good indicator of instrumental value, however, nor is instrumental value a good indicator of biological integrity. Nevertheless, I believe that we should care about the health and integrity of biological communities, even if these qualities bear no general logical or empirical relationship with the instrumental value of those communities.

Two Kinds of Ecological Science

A few years ago, Charles H. Peterson (1984:129–130) described a "general philosophical schism (among ecologists) separating those who believe that nature can be improved by the works of man (the 'biotechnologists') from those who treasure and seek to preserve the biological and ecological status quo (the 'bioconservatives')." Among ecologists, bioconservatives value nature for noninstrumental reasons and think of biological communities as being good in themselves or as having a good of their own. Accordingly, they resist changes that may improve the instrumental value but damage the "health" or "integrity" of these systems.

The biotechnologists, in contrast, think of systems and communities as existing for our good rather than as having a good of their own. They propose to show "policymakers how natural ecosystems might be better used for man's well-being" (Dohan 1977:134). Accordingly, they would change the design of these systems or replace them altogether to increase the long-run returns to humanity as measured by goods and services for which people are willing to pay (Van Dyne 1969).

Dr. Frank Forcella, writing in the *Bulletin of the Ecological Society of America*, recently called on his ecologist colleagues to embrace the biotechnologist credo—"to engineer and produce plants, animals, and microbes that better suit the presumed needs and aspirations of the human population." Forcella continues:

> Ecologists are the people most fit to develop the conceptual directions of biotechnology. We are the ones who should have the best idea as to what successful plants and animals should look like and how they should behave, both individually and collectively. Armed with such expertise, are we going to continue investing nearly all our talents in Natural History? Or should we take the forefront in biotechnology, and provide the rationale for choosing species, traits, and processes to be engineered? I suspect this latter

approach will be more profitable for the world at large as well as for ourselves. [Forcella 1984:434]

Forcella is correct both in the way he divides ecologists into two groups and in the conclusion he draws from that division. One group of ecologists—those who study the effects of evolutionary history—seek to account for the abundance and distribution of species. They give us explanations that appeal to contingent and random events; theirs is a historical science. These ecologists may argue, moreover, that Forcella's approach ignores everything that is spontaneous, free, and contingent in nature—everything that makes nature an appropriate object of wonder, moral attention, and appreciation. They may agree with Stephen Jay Gould that

> with contingency, we are drawn in; we become involved; we share the pain of triumph or tragedy. When we realize that the outcome did not have to be, that any alteration in any step along the way would have unleashed a cascade down a different channel, we grasp the causal power of individual events. We can argue, lament, or exult over each detail—because each holds the power of trans-formation. [Gould 1989:284]

It is not a coincidence that many of the best observers of nature—Audubon would be an example—were artists as well as scientists. Ecology may differ from painting less in its purpose or even in its methodology than in the symbols it uses and in the questions it asks. Ecology, like poetry and art, may help us preceive and appreciate what is wonderful in nature and, therefore, make us more willing to protect magnificent ecosystems for their own sake.

An opposing school of ecologists seeks to determine the principles and techniques by which we can manipulate, control, and predict ecological events. This kind of ecology, rather than explaining the story of the past, seeks to design the communities of the future. These ecologists, as Dr. Forcella suggests, investigate the principles and techniques by which we can improve plants and animals or replace them with more profitable biological machines and systems. After all, the blind forces of evolution—random mutation and local natural selection—could hardly be expected to keep up with our changing consumer preferences.

One may make the same division in other sciences. Consider the study of minerals. Geologists study the physical history of the earth. They help us to account for the abundance and distribution of minerals. Metallurgists, in contrast, pursue the techniques and the science of making and compounding alloys. No one would ask a metallurgist to explain why there is more iron in Michigan than in Massachusetts. Similarly, no one would ask a geologist to create an alloy for a new razor.

Consider another science: psychology. Many psychologists— those who are concerned with market research, for example— study how to control, predict, and manipulate human behavior to deliver votes, consumer dollars, or whatever to those willing to pay for this service. These behavioral psychologists, using methods that John Watson and B. F. Skinner pioneered, possess an increasing expertise people can use to get other people to do what they want. This approach in psychology, as Skinner rightly pointed out, takes us beyond freedom and dignity into the realm of the purely instrumental.

Many other psychologists—therapists, for example—concern themselves with improving the mental health or well-being of their clients. They often study the personal history of their patients to help them understand their behavior in the light of that history. In a therapeutic context, the freedom and dignity of the patient count. He or she is seen not as an object of use but of love and appreciation.

Plainly, notions of "integrity" and "health" would have no role in the practice of the first sort of psychological science, but they would be central to the practice of the second. This is true because the former discipline seeks to manipulate and control individuals for economic or other purposes, while the second concerns itself with the good of these individuals in themselves. This is also true of ecological science. Notions of the "health" and "integrity" of natural communities would be useless to those who wish to design the factory farms, put-and-take fisheries, and aquacultural feedlots of the future. Only those who consider nature an object for moral and aesthetic attention need care about the integrity and health of ecosystems.

Dr. Forcella observes correctly that ecologists who study natural history and who concern themselves with the integrity of ecological communities miss a great opportunity to profit both themselves and the rest of humanity. They miss the opportunity to replace naturally occurring plants and animals with domesticated ones (like the seedless grape and the self-basting turkey) that are plainly far more marketable. And they miss the chance to design industrial methods of bioproduction, biosubstitution, and biorelocation (like those that allow the manufacture of cocoa butter from inexpensive vegetable oil) to spew out plant and animal tissue at increasing rates and decreasing prices.

Consider, for example, recreational fisheries. It is absurd to think of hatchery-based put-and-take fisheries as having a "good" of their own. Yet they satisfy the consumer demands of the fishing public. Diverse, complex, resilient ecological communities—communities that exemplify "health" and "integrity"—cannot produce the catches that millions of fishermen want. This is true because put-and-take fisheries are based on stupid fish raised in captivity that will gobble up anything, thereby rewarding even the most inept angler with a good return. Managers know that the majority of consumers are willing to pay for this return. To maintain the integrity of natural ecological systems, in contrast, managers have to put strict limits on what people can catch. And anglers would have to develop the patience and love of nature that relate to the intrinsic value of fishing but not to the demands for which most consumers of wildlife resources appear willing to pay.

Ecologists may help us develop the Chesapeake Bay not only as a site for hatchery-based sure-catch fishing, but also into a more efficient sewer, liquid highway, and aquacultural runway. They may work with biotechnologists to design vast industrial fermentation systems in which plant cells, engineered to reproduce like microorganisms, will multiply endlessly to produce the agricultural surpluses of the future. They might eventually fabricate something close to Al Capp's schmoo, an animal that services all our needs, thrives under any conditions, and makes no demands. We might then consign our evolutionary and ecological heritage to the dustbin of obsolescence. Life itself would become an industrial commodity—the result of economic but not biological competition.

Ecology as a Predictive Science

Those like Frank Forcella who believe that ecology should help us domesticate nature and eventually replace its products with economically more valuable ones might offer not only instrumental reasons but also the following methodological argument. Domesticated, engineered, or artificial organisms and ecosystems serve better than natural ones as laboratories for discovering and testing general principles of biological science. This is because natural history defies mathematically precise testable generalizations, while artificial systems obey fairly simple input/output models. The population biology of whales remains a mystery, for example, but Mr. Perdue can tell you virtually to a feather how many chickens he will bring to market.

Ecologists who create artificial species and ecosystems, then, may have an advantage over those who study natural communities—namely, the ability to control and therefore to predict events. Advances in biotechnology now make it possible for ecologists to shape the very species and ecosystems they study. Rather than molding their models to fit the world, they mold the world to fit their models. Since they test general principles and hypotheses about the living world, they may be in a position, moreover, to answer calls for a "predictive" ecology. Accordingly, ecologists who take the forefront in biotechnology and bioengineering may discover "laws of ecology" that elude those who study natural history.

If being "scientific" means being predictive rather than descriptive, ecological science should focus on the behavior of artificial not natural biological systems. In other words, if the methodology of science requires strong inference from "laws of nature," ecologists, insofar as they are scientists, should stop studying natural ecosystems altogether. They should primarily investigate man-made ecosystems—sewage treatment plants and factory farms, for example, about which testable, universal, and mathematically precise knowledge appears attainable. "In agricultural ecology," as J. L. Harper points out, "precision, realism and generalization become more compatible in scientific inquiry. It may well be for this reason that much of the development of ecology as a science in the post-descriptive phase will come from the study of man-managed ecosystems" (Harper 1982:22).

Integrity, Productivity, and Scale Effects

Why, then, would ecologists take any interest in natural history? Why should they think or care about the "health" and "integrity" of biological systems?

In congressional testimony in 1971, Richard Carpenter suggested one answer to this question. "Ecological principles," he said, "show that, for the long term, maximum productivity coincides with a healthy, esthetically pleasing environment" (Carpenter 1971:5). Carpenter implies that the choice between use value and intrinsic value arises only in the short run. In the long term, policies that protect the intrinsically valuable qualities of nature will enhance its instrumental value as well.

The argument that "maximum productivity coincides with a healthy, esthetically pleasing environment" makes sense when we compare healthy ecological systems with those that have been polluted or degraded in capricious and criminal ways—superfund sites, for example. But nobody defends careless and criminal practices. The judgment we must make compares the long-run productivity of technology-based bioindustrial systems associated, for example, with agribusiness and the productivity of ecosystems resplendent with complexity, resilience, diversity, homeostasis, hierarchy, and other qualities associated with integrity and health.

Which is more productive from an instrumental point of view—the tallgrass prairie or amber waves of grain? In the prairies and great plains complex communities of native flora and fauna have given way almost entirely to monocultures of non-indigenous domesticated wheat and corn and to the agroindustrial systems that support them. The tallgrass prairie exemplifies resilience, complexity, diversity, and other qualities associated with the health and integrity of ecosystems. Yet when compared to agribusiness, the tallgrass prairie may score high with respect to intrinsic but not instrumental value.

Someone may reply that a farm using high-input and high-tech methods will count as "productive" only if productivity is measured in terms of yield-per-acre. But it may be less productive in terms of yield-per-calorie of energy expended or yield-per-unit of soil eroded or yield-per-pound of pesticide or fertilizer used. The fuel burned, the soil eroded, and the pesticides poured, moreover, have long-term ill effects that may cancel out apparent productivity gains. Thus, the instrumental value of a system may depend on questions of scale—the extent to which we trace relationships in time and space—rather than on just the short-run return on investments.

After careful study of current agricultural practices, however, a prominent team of experts at Resources for the Future (RFF) predicts that surpluses will continue to beset commodity markets as far into the future as one can see. Using a social discount rate and other techniques of microeconomic analysis, many resource economists argue that erosion, for example, poses little threat to productivity. The leader of the RFF study acknowledges, though, that if people care about the land for aesthetic and moral reasons, they will—and perhaps should—turn a deaf ear to evidence that we do not need the good earth for instrumental purposes. "If the concern with the adequacy of agricultural land is really a concern for amenity values," Pierre Crosson concludes, "even the strongest arguments that land preservation is unnecessary to maintain productive capacity will not be persuasive" (Crosson 1983:191).

The "externalities" of industrial production that most concern us, however, have to do with the deterioration of the planet's biospheric support systems. We worry, therefore, not so much about the "external" effects one economic activity may impose on another (for example, efficiency in the allocation of resources) as about the effects of the scale of economic activity as a whole on the planet's ecological carrying capacity. As economist Herman Daly puts this point: "Optimal allocation of a given scale of resource flow within the economy is one thing (a microeconomic problem). Optimal scale of the whole economy relative to the ecosystem is an entirely different problem," a macroeconomic problem (Daly 1991:35). Daly offers the analogy of a boat that sits lower and lower in the water as it takes on more cargo. When the water comes to the Plimsoll line, the boat has reached its carrying capacity. Even if the weight in the boat is optimally or efficiently allocated, the boat will still sink if it receives more cargo. "It should be clear," concludes Daly, "that optimal allocation and optimal scale are quite distinct problems" (p.46).

While Daly's argument is clear and convincing, it primarily concerns the integrity of global biospheric processes, not that of local ecological communities. It tells us how to adjust economic activity as a whole to these larger processes—by recycling wastes, controlling carbon emissions, and, in general, limiting the amount of low-entropy resources we take from nature and high-entropy wastes we put back in. Daly's argument, in other words, applies to the adjustment of macroeconomic to macroecological processes. It has no clear implications for the health, integrity, and value of local natural communities and environments.

The particular global threats that most concern us—for example, the depletion of the ozone layer and the greenhouse effect—arise from the release of pollutants primarily from industry directly into the atmosphere. Even if the Chesapeake Bay remained in the pristine condition in which Captain John Smith found it, this would have no effect on those problems. Global biospheric processes may be linked to local ecological communities; for example, the eutrophication of the bay may store carbon and thus counter global warming. The relation between the "health" and "integrity" of local systems and global processes may be positive or negative, or may be a mixture of both. The connection between global problems and local ecological changes—for example, the replacement of climax ecosystems by agricultural crops that add net biomass and thus absorb net carbon—is a complicated one to which few, if any, generalizations may apply.

ECOLOGICAL HEALTH AND INTRINSIC VALUE

If we are to accept the idea that ecosystems may be objects not only of use but also of aesthetic appreciation and moral attention, then we must accept the possibility that these systems have a good of their own we ought to respect and therefore protect. If they are living systems that have a good of their own, we are able to ascribe states of health to them—or, if "health" is a privative concept (as it may be), then at least states of illness and disease.

Concepts of ecological health and integrity and their deployment in ecological science, as I shall argue, make the most sense in relation to the intrinsic—the moral and aesthetic—value of ecological communities and systems. To suppose that natural communities may have intrinsic and not just instrumental value, after all, is to distinguish what may be healthy for nature from what may be healthful or beneficial for humanity. Ecologists who take a bioconservative position—who wish to speak intelligibly of the intrinsic value of nature—must therefore develop a concept of ecological health that is not simply a function of the welfare of human beings.

To introduce notions of "health" and "integrity" into their science, ecologists will need to adopt a bioconservative position; in other words, they will view nature as an object of aesthetic appreciation and moral respect and not simply as a source of human benefits. This does not entail, however, that the integrity of ecological systems requires they remain pristine or protected absolutely from human use. Such a criterion—historical authenticity—if used as the measure of ecological health, would doom that concept to irrelevance.

This is true, first, because the impact of culture and cultivation on nature has been so pervasive and endemic that nature may be impossible to identify—and it is certainly impossible to restore or recover—as it would have been in the absence of human beings. The people who crossed the Bering Straits and killed off the woolly mammoth and saber toothed tiger probably already did the job of completely changing the course of natural history in the Americas. An attempt to tie the integrity of environments exclusively to their evolutionary history may represent an empty exercise of nostalgia. In many instances, the biological status quo we may wish to protect may have resulted from human activity a generation or two earlier. Species that seem most at home in the United States today—Kentucky bluegrass, for example—hitchhiked here with human colonists. Is bluegrass "authentic" or not? Is it healthful or damaging to local ecosystems? Even conceptions of degree—of relatively more or less "pristine" environments—may be impossible to define in ecological terms.

Second, many ecosystems whose character has been changed radically by human activity—the Lake District in England, for example—may appear quite "healthy," and they surely attract our admiration, respect, and moral attention, in spite of and sometimes because of those changes. Accordingly, evolutionary authenticity (the condition of being unaffected by human use) cannot be a necessary condition for ecological integrity.

Third, we cannot avoid interventions into nature. Even wilderness areas must be managed; we "play God in Yellowstone" as well as everywhere else. What is more, economic and political realities—in Brazil, for example—compel us to integrate economic use with environmental protection and thus to manipulate ecosystems in ways that do not threaten their health and integrity.

Hence, we cannot usefully define ecological "health" or "integrity" simply in terms of ecological "authenticity"—that is, in terms of the independence of natural from cultural history. We must develop a conception of health that is compatible with the cultivation and use of nature but that does not collapse into the maximization of instrumental value. The task of developing nature as a habitat for human beings has been the traditional work of human culture. The word "culture" derives from *colere*—to cultivate, to dwell, to care for, and to preserve—and relates primarily, as Hannah Arendt notes, "to the intercourse of man with nature in the sense of cultivating and tending nature until it becomes fit for habitation. As such, it indicates an attitude of loving care and stands in sharp contrast to all efforts to subject nature to the domination of man " (Arendt 1968:211–212).

The attitude that Arendt recommends may require us to conceive of nature not just as a collection of resources or materials for our use but as the habitat—essentially, the place—in which we live. By preserving the health and integrity of these species and communities, we ourselves sink our roots in the land we inhabit: we ourselves become native to a place.

CONCLUSION

William Reilly, administrator of the Environmental Protection Agency, recently wrote: "Natural ecosystems have intrinsic values independent of human use that are worthy of protection." He cited an advisory scientific report that urged the agency to attach as much importance to intrinsic ecological values as to risks to human health and welfare. Reilly added:

> Whether it is Long Island Sound or Puget Sound, San Francisco Bay or the Chesapeake, the Gulf of Mexico or the Arctic tundra, it is time to get serious about protecting what we love. Clearly we do love our great water bodies: We flock to them to live, to work, and to play. They are part of our heritage, part of our consciousness. Let us vow not to let their glory pass from this good Earth [Reilly 1990:4].

In 1991, the State of Maryland offered anyone registering an automobile the option of paying $20 (which would go to an environmental fund) to receive a special license plate bearing the motto: "Treasure the Chesapeake." A surprising number of registrants bought the plate. How many of us would have ponied up the $20 for a plate that read: "Use the Chesapeake Efficiently" or "The Chesapeake: It Satisfies Your Revealed and Expressed Preferences?"

We treasure the Chesapeake—and we cherish other environments—because we ourselves are native to them. The value we attach to them goes to our identity more than to our interests—to who we are, not just what we want. In that sense, the health and integrity of ecosystems becomes a necessary condition of our own health and integrity. If we fail to attend to the health and integrity of local environments, in other words, we will ourselves lose our moorings; we will become as standardized and fungible as the pigs that go to slaughter. And our identity, too, will depend on the nature of capital and technology rather than on the nature that gives us life.

To treasure an ecological community is to see that it has a good of its own—and therefore a "health" or "integrity"—that we should protect even when to do so does not profit us. "Why did you do all this for me?" Wilbur asked. "I've never done anything for you." Even when nature does not do anything for us—one might think, for example, of the eagles and otters destroyed in Prince William Sound—we owe it protection for moral and aesthetic reasons. Otherwise our civilization and our lives will amount to little more than what Charlotte described as "all this trapping and eating flies."

In this essay, I have proposed that we may lift up our lives a little by seeing nature as Charlotte did, not just as an assortment of resources to be managed and consumed, but also as a setting for collective moral and aesthetic judgment. In this setting, the concepts of ecological integrity and health make sense. Our evolutionary heritage—the diversity of species, the miracle of life—confronts us with the choice Farmer Zuckerman had to make: whether to butcher nature for the market or to protect its health and integrity as objects of moral attention and aesthetic worth.

If Zuckerman had not learned to appreciate Wilbur for his own sake, he would have converted the pig to bacon and chops. Likewise, if we do not value nature for ethical and aesthetic reasons, then we might well pollute and degrade it for instrumental ones. If a spider could treat a pig as a friend, however, then we should be able to treat a forest, an estuary, or any other living system with affection and respect.

References

Arendt, H. 1968. *Between Past and Future: Eight Exercises in Political Thought.* New York: Viking Press.

Carpenter, R. 1971. "Goals and Policies for Environmental Improvement." In *Congress and the Nation's Environment.* Report prepared by the Environmental Policy Division, Library of Congress, for the U.S. Senate, Committee on Interior and Insular Affairs, 91st Congress.

Crosson, P. 1983. "Future Economic and Environmental Costs of Agricultural Land." In P. Crosson (ed.), *The Cropland Crisis: Myth or Reality?* Baltimore: Johns Hopkins University Press.

Daly, H. E. 1991. "Elements of Environmental Macroeconomics." In R. Costanza (ed.), *Ecological Economics: The Science and Management of Sustainability.* New York: Columbia University Press.

Dohan, M. R. 1977. "Economic Values and Natural Ecosystems." In C.A.S. Hall and J. W. Day (eds.), *Ecosystem Modeling in Theory and Practice.* New York: Wiley.

Forcella, F. 1984. "Commentary: Ecological Biotechnology." *Bulletin of the Ecological Society of America* 65:434.

Gould, S. J. 1989. *Wonderful Life: The Burgess Shale and the Natural History.* New York: Norton.

Harper, J. L. 1982. "After Description." In E. Newman (ed.), *The Plant Community as a Working Mechanism.* Oxford: Blackwell.

Peterson, C. H. 1984. "Biological Consequences of Manipulating Sediment Delivery to the Estuary: A Blueprint for Research." In B. J. Copeland et al. (eds.), *Research for Managing the Nation's Estuaries.* Publication UNC-84-08. Raleigh, N.C.: UNC Sea Grant Program.

Reilly, W. K. 1990. "A Strategy to Save the Great Water Bodies." *EPA Journal* 16:(6):4.

Van Dyne, G. M. 1969. *The Ecosystem Concept in Natural Resource Management.* New York: Academic Press.

White, E. B. 1952. *Charlotte's Web.* New York: Harper & Row.

4

Toward an Open Future:
Ignorance, Novelty, and Evolution

MALTE FABER, REINER MANSTETTEN,
AND JOHN PROOPS

Economic activities in the industrialized world are placing more and more pressure on ecological systems. Indeed, modern economies tend to cause stress, illness, and destruction to the ecological systems in which they are embedded. This situation confronts us with problems that are not only very pervasive but also very unusual. To come to grips with these problems we will use *openness* as the key concept. While the notion of open systems is already well known in the literature, we introduce the notion of the *openness of the future* to this debate. Our purpose here is to develop an evolutionary conceptual framework, relating the openness of the future to our response to the ecological impact of human activity, using such concepts as surprise, novelty, chaos, and ignorance. We apply this conceptualization to problems in environmental policy.

In order to explore the consequences of an important generalization—that we always act in open-ended situations—for the interaction of human economic systems and natural ecological systems, we begin by asking: What attitude has led humankind to the continued endangering of nature and thus of its livelihood? We begin our investigation by characterizing the driving force of this attitude—the attempt to create a human world that is closed against the influence of uncontrollable nature—in general terms. We then offer a conceptualization of closure and openness within an evolutionary framework in order to develop a scientific basis for analyzing the notions of novelty and ignorance and to explain their relevance for what we can know, what we can control, and what we can do. We next outline some of the policy consequences of a new attitude toward closure and openness. In particular, we show that because of the different types of novelty they generate, the problems of resource use will tend to be less difficult to address than the problems of pollution. Finally, we explain some of the implications of our analysis for concerns about ecosystem health and integrity.

THE TWO HOUSES OF HUMANKIND

Within the name "ecological economics" we find the Greek word *oikos* twice. The meaning of *oikos* is house. Ever since Aristotle, economics has been the science of the oikonomia—the science of all problems concerning the allocation and distribution of labor, goods, and services within the *oikos*. Here *oikos* can mean the house of the family as well as a town or even a state. The *oikos* in the words economy and economics is clearly a house whose founder and master is humankind.

The word "ecological," however, draws our attention to the circumstance that we, as human beings, live together with animals and plants in a common house, *oikos*, that comprises the whole living space. This *oikos* has a certain internal order, which is responsible for the development and evolution of life within it; this order is indicated by the suffix "-logical," which is derived from the Greek word *logos*. The literal meaning of *logos* is "word," but it may also be translated as "concept" or "structure." In the philosophy of Heraclitus (sixth century b.c.), logos is the principle of all beings and thus of the whole universe.[1]

The concept of logos is useful to formulate a broad definition of ecology: Ecology is the science of the principles of the self-organization of nature. According to these principles new orders, ecosystems, species, plants, materials, and so forth, evolve and old ones are overcome or destroyed. The embodiment of these principles we will call logos. For Plato and Aristotle, as well as philosophers of many religions, the assumption of a logos or similar concepts—whether derived from a divine creator or a principle of nature itself—was a prerequisite for the true understanding of nature. We believe that any adequate theory of the management of economic systems that also protects ecological health, must rest on a clear recognition that the structure of ecological systems is a function of their ability to organize themselves.

It is evident that the house of nature and its principles were not founded by humankind; we are guests in this house, as are all other living beings. In modern industrial times, however, it appears that humankind has claimed the common house as its property. In many respects, humankind behaves as if the common house, with all the inhabitants, was founded only for its use. It appears, therefore, that modern humanity has conceived the common house of nature as a part of the human house. In contrast to this view, in an ecological perspective the human house (the economy) is part of the house of nature. This perspective is still rather new in the attitude of modern societies, and we have to admit that the West up to now has not made much effort to understand the logos of nature's house.[2]

During the last two decades, however, humankind has increasingly become aware that it has changed the order of its common house. Therefore, it seems appropriate to take a fresh look at the concept of nature's logos. Humankind will be unable to live in accordance with nature without acknowledging its logos, and hence without knowing how it influences and changes its common house and what consequences its actions have.

The main question ecological economists are confronted with is this: Can we live in our human house, in a modern economy with its intrinsic dynamics, and at the same time respect nature's own logos? And if we can, how can we achieve that? These tasks have also been formulated by Paul Ehrlich (1989:14): "Somehow a new ecological-economic paradigm must be constructed that unites (as the common origin of the words ecology and economics implies) nature's housekeeping (i.e. the logos of nature, the authors) and society's housekeeping, and makes clear that the first priority must be given to keeping nature's house in order." We turn now to some prerequisites to deal with this encompassing task.

CLOSURE AND OPENNESS

Our first approach to describing the attitude of modern humankind is guided not by science but by poetry, namely by the last act of Goethe's *Faust* (see Binswanger, Faber, and Manstetten 1990). Let us imagine a rather infertile stretch of land near the sea. A few people earn their living by fishing, raising livestock, and a little agriculture. Their economy is self-sustaining and stationary. After coming into being the economy does not affect the ecosystem any more. The life of the people is hard: fishermen sometimes die at sea, famines and epidemics come and go. Let us further imagine that in this archetype of a premodern life-style, there arrive entrepreneurs and engineers. They see that this stretch of land is utilized in an inefficient way. They propose to restrain the sea by constructing a system of dikes, and then to drain and cultivate the land. This will enable them to erect towns and industries: a living space for millions of people can be secured. Of course, this is only possible if the former inhabitants give up their way of life and take over the new life-styles of the entrepreneurs and engineers. The latter view is the view of Faust, the hero of Goethe's drama.

The dikes may be conceived of as a symbol for the Faustian intention to separate the house of humankind (economy) from the house of nature (ecosystem). The dikes are the walls of the house of humankind—and nature is allowed to enter only if it is necessary or useful for human purposes. In all other cases the house of humankind has to be kept closed against the uncontrolled forces of nature, which in *Faust* are symbolized by the wild sea. But this uncontrolled nature, seemingly kept outside, presses against the dikes day and night. Since even one hole may have disastrous consequences, the dikes must always be supervised. Hence there exists an enduring danger: "But Care through the key-hole slips stealthily in" (*Faust*, Part II, Act 5). But as long as the human house can be kept closed, in the way described here, humans can suppress their cares and anxieties and lead their lives according to their desires. Thus humans gain safety, sanity, and welfare: they are able to pursue their own happiness. All developments in the house of humankind seem to depend only on humans themselves and on nature in a narrow sense: a stock of resources for the fulfillment of human wishes. The only "logos" of this kind of nature is derived from human will and consists in its utility for human intentions.[3]

It is important, however, to recognize that the house of humankind is built without considering nature's own logos. Now nature outside the house of humankind is experienced as having no sense in itself and as being potentially dangerous. Thus much intellectual and psychic effort, as well as many resources and much energy, are needed to keep the human house closed against those aspects of nature not corresponding to human intentions. Having described how nature's logos is excluded from the house of Faustian humankind, we now turn to the perception of time.

The Faustian Future

The dominating force in this Faustian world is the will of humankind. It is the will to be, as far as possible, the master of everything that can happen. This impulse will also deeply affect the treatment of time. When the approach to the world is guided by the Faustian will, the perception of time does not take regard of the past. The past seems to be nothing but a presently accessible stock of information and material of all kinds (resources, capital goods, consumption goods, and so forth). These can be utilized to control whatever is encountered. It is in this sense that "time is reduced to only one function, namely to the future as the space where economic growth takes place." (see Binswanger, Faber, and Manstetten, 1990, for a more detailed explanation.) There are three kinds of future events with which Faustian humankind has to deal:

- Events that seem to be dangerous; their effects must be neutralized.
- Events that are advantageous; their effects must be utilized.
- Events that are neither dangerous nor advantageous; their effects can be neglected.

In this Faustian world, humankind is convinced that if human control of events can be achieved and maintained, the future will be characterized by progress and increasing welfare. This implies the expectation that humanity will be increasingly able to solve all problems in the future. Keynes indeed believed that humankind can solve its problems by its own efforts, although he admitted this may be difficult: "The pace at which we can reach our destination of economic bliss will be governed by four things—our power to control population, our determination to avoid wars and civil dissensions, our willingness to entrust science to the direction of those matters which are properly the concern of science, and the rate of accumulation as fixed by the margin between our production and our consumption; of which the last looks easily after itself, given the first three." (See Keynes 1963:373; see also in a similar vein Solow 1973:42.)

The creative mind in the Faustian world allows an inventiveness that seems to guarantee infinite progress. The future, and thus time, is open—but only to progress. To understand the consequences of the attitude that has led to the tripartite categorization of future events, we shall look for its implicit premises. This attitude can only be sustained if all future events that concern the interest of humankind are either predictable or immediately controllable; if the prediction can be made sufficiently in advance of the event; and if there are techniques available to control these events in accordance with the interests of humankind.

If these premises are fulfilled, humanity is able to control everything that concerns it. In this respect, humankind may regard itself as godlike, being at least omniscient and omnipotent. (The power and pervasiveness of modern telecommunications may even give the illusion of omnipresence; that is, the abolition of space.) The Faustian illusion of omnipotence itself requires the illusion of omniscience, for if everything is to be controlled, then everything must be known. We therefore see that the Faustian will to control excludes the possibility of ignorance, either in the present or in the future. The assumption of perfect knowledge about the future itself presupposes that there can be no emergence of novelty in the future.

Traditional Science and the Future

The Faustian illusion of omniscience and omnipotence implies that nature is, in essence, predictable. Modern scientific method, deriving as it does from a society dominated by the Faustian imperative to control, quite naturally takes the predictability of the world as its cornerstone. Scientific method therefore includes among its intellectual and social functions the following: the enlargement of the scope of predictability, the enlargement of the period of time over which predictions may be made, and the development of technologies to deal with the predicted events. We shall concentrate here on the first aspect, which is of fundamental importance for the second and third.

When scientists analyze a natural or social process, they attempt to understand it in such a way that the whole structure of the corresponding process and all its interdependencies can be explained by scientific laws. In the ideal scientific case, this structure is causal—that is, any state of the process can uniquely be conceived as the effect of earlier states (and the cause of future states) of the process. This implies that at any given point in time, we can reconstruct all past states and predict exactly all future states. This reconstruction is only possible if there exists no unwanted novelty and there is no ignorance.

If, in this way, we could successfully examine all processes in the world, and all their interdependencies, and could know the complete state of the world at any time, then it would, in principle, be possible to compute all the states of the world at any moment in the future or the past. This situation corresponds to that of Laplace's demon. (See Prigogine and Stengers 1984; Faber, Niemes, and Stephan 1987:91.) In a world governed by the Laplacian ideal, we are interested to know about as many natural and social processes as possible. For if we could predict the future development of such processes and if we knew their initial conditions, then we would be able, by influencing the initial conditions, to determine the outcome of these processes, and therefore to control the future.

Hence science, and its application in the form of technology, promises the increasing realization of human wishes and intentions and is therefore a prerequisite for the closure of the house of humankind. Thus, for example, we can influence the yield of a crop by employing specially bred seeds, artificial irrigation, chemical fertilizers, and pesticides. Such an example shows that science enables us, under certain circumstances, to become secure against certain unwanted future developments such as famine and drought. Positively expressed: Science and technology enlarge the set of feasible possibilities and therefore the scope of freedom of action. They are the basis of the Faustian world described earlier.

While in the nineteenth century Laplace's demon was an ideal for natural scientists and became one of the driving forces for society at large, we know now that this ideal cannot, even in principle, be reached. This implies that we have to accept that the future is, in principle, open: we cannot know everything that will happen. Thus the Faustian project of completely closing the house of humankind to the house of nature cannot, in principle, be fully achieved. (See Faber and Proops 1990:chap. 2 and 5.)

Welfare and Security vs. Hubris and Anxiety

What are the consequences of the Faustian attitude to controlling the future? In those parts of the world where humankind has achieved considerable control over many natural—and to some extent social—processes, it has enlarged its possibilities of action to a tremendous degree. The dynamics of this development are amazing even in retrospect. This has resulted, at least in some parts of the world, in a vast increase in security and welfare in three respects: protection against disasters and disease; the provision of plenty of food; and the delegation of much human labor to machines. Thus many humans can now do what they want to do, rather than what they must do.

The experience that many important processes can be controlled has led to the belief that, in principle, all processes can be submitted to human management by means of science and technology. This attitude of overweening pride is the hubris of the Greek myths; for hubris means that humanity loses perspective and feels itself to be godlike.

Although this belief in progress no longer exists in this extreme form, there are still many traits and relics of it at work. In the 1960s, for example, the Kissimmee River in Florida was channelized by a huge Army Corps of Engineers project. The loss of wetlands habitat was disastrous, as were the downstream effects on Lake Okeechobee. The state of Florida is now exploring the possibility of restoring the meandering river, but the cost of partially restoring lost functions threatens to be 10 times the original cost of the channelization.

A further example: during the 1960s and 1970s in the Federal Republic of Germany it was supposed by industry, by government, and by great parts of the public that nuclear power could be governed and that the wastes from nuclear power plants could be safely disposed of or treated in processing plants. Hubris easily gives rise to anxiety, because humankind feels that there always exists the possibility, even if the probability is very small, that for one or the other reason one might lose control. The example of nuclear power illustrates also that hubris and anxiety may be met in different groups of society regarding the same area of concern. Outside the nuclear industry, there already existed in the 1970s an articulate minority that expressed great anxiety because of the uncontrollability of the nuclear industry.[4]

Here we can recognize a temporal pattern. First, when the nuclear program was decided upon, hubris predominated. This also held true for the first decade of its realization. But the more nuclear power plants that began operation, the more people became aware of possible dangers of the nuclear technology and the more anxieties arose. While hubris arises from an overestimation of our abilities to control technology and its impact on natural processes, in a state of anxiety we become less confident in our abilities to react and eventually to avoid the dangers. Of course, hubris and anxiety are exhibited, as a rule, together within one human being, though they show up in different manifestations. Thus the president of a firm may give the impression that everything is under control during the day, while in the evening he tells his wife about his anxiety.

Greek Philosophy vs. Modern Science

Here it may help to contrast the attitude of modern humankind with holistic ways of understanding the world. Such ways were developed in ancient India, China, Greece, and elsewhere. Plato, in his philosophy of nature, offers a view of the world that seeks not to influence natural processes but simply to contemplate them in their eternal "idea" and in this way to experience their essence. Thus the Greek word *theoria* means, literally translated, "contemplation" (see Faber and Manstetten 1988:100). Plato (1953:716ff; 28aff), like many other Greek philosophers, conceived the whole world as one sentient being with the "finest body" and the "most noble soul."

All processes, be they mental, organic, or inorganic, express the life of this sentient being, the "cosmos." The latter, in turn, is an image of the divine life itself. (Compare the Gaia concept in Lovelock 1979.) According to this view of the world, it is the task of all *theoria* to recognize the right place of all things and beings in the great life of the world. This holds also for human beings. Having found one's place, a conscious being must act according to its insight into this great life of the world. In this view, life may unfold itself anew in each moment. The future is open for fortune as well as misfortune. The true happiness of human beings consists in this openness concerning the unfolding of the cosmos. From this it follows that the notion of predictability was not important for Greek philosophy.[5]

In contrast to Greek science, the measuring rod of modern science is not the great life of the cosmos. Such an idea has no meaning, for example, in classical mechanics. Instead science is oriented to improving the conditions of human life, which implies that natural processes are influenced and controlled as much as possible (see Faber and Manstetten 1988:102ff), but the modern, Newtonian paradigm postulates a deterministic world that is unattainable.

First Summary

The ecological crisis has led to different reactions by different groups of people. A prominent reaction is the claim for increasing one's efforts to control the economy by technological and political means in order to protect the environment. Sometimes there is even a demand for an "eco-dictatorship." In our eyes, such actions remain Faustian in attitude. Clearly we cannot avoid employing Faustian measures to a certain extent. But we doubt that it will be sufficient to rely solely on this kind of measure, for it is our Faustian spirit that has led to the present situation. Faustian will and science will not solve our problems. Therefore, we consider it essential that the spirit in which necessary measures are taken must broaden toward openness.

In the following section we will try to develop a concept of openness by conceptualizing the notions of surprise, ignorance, novelty, and evolution. For we believe that the environmental crisis not only consists of dangers but also opens new ways of living and thinking. It offers us the chance to transcend the concepts and attitudes that block us from access to more openness and thus to the true fullness of life. Nature is sufficiently resilient to incorporate human impacts, provided we do not destroy its logos—its capacity for self-organization and creative reaction to novel events. Therefore we believe that the embodiment of the principles of evolution—the logos of the house of nature (in Ehrlich's terminology: nature's housekeeping)—will remain intact despite destructive human activities. Clearly this confidence is not based on science: it is a faith, a faith in the logos of nature. This faith implies that humankind cannot essentially interfere, despite all its actions, with the very logos of nature.

Of course, such a faith does not correspond to any truth that can be proved or demonstrated scientifically. This very principle of the evolution of the universe shows itself in the emergence of novel solutions in a dynamic system; however, the latter cannot be known ex ante. Thus the logos is beyond our knowledge; it corresponds rather to our ignorance. Hence we cannot know if and how evolution develops, but we may hope that, in spite of human-made ecological damage, the principle of evolution may generate new forms of life and perhaps of intelligence. While our knowledge might suggest that humanity is going to destroy the ecological basis for its survival, the acceptance of our ignorance makes us open to new, heretofore unknown, possibilities that may allow humankind to survive. Therefore, the faith in the logos of nature does not require any theological and philosophical dogmatism, but it has, as a consequence, an attitude of openness.

Even if we accept that the logos of nature is not a direct object of science, we will, of course, try to learn as much as possible about nature's principles. This endeavor will be furthered by an attitude of openness and an experimental attitude toward living within natural systems, even as we partially control them. This in turn will help us to accept that all our knowledge is related to a fundamental ignorance. While the Faustian attitude causes us to fasten our grip on the economy and the control of nature, be it for exploitation or for environmental protection, the attitude of openness (the acknowledgment of ignorance as well as the readiness to discover and learn whatever is necessary) allows humankind to let loose and to hold fast at the same time.

CONCEPTUALIZING OPENNESS

A key characteristic of the fullness of life is surprise. Our everyday life is full of surprises: we experience pleasant or unpleasant situations and events that we would never have expected. On the one hand, in a Faustian society people try to insure themselves against unwanted surprise—against suffering and unpleasantness. On the other hand, if there is not enough surprise experienced in everyday life, people tend to create it in various ways—they practice risky sports, for example, or become involved in love affairs. In contrast to our everyday life, the way science has conceived and handled its tasks up to now depends heavily on the attempt to exclude surprise.

Surprise, Ignorance, and Openness

What does surprise mean? We conceive surprise to be related to human consciousness, to knowledge, and in particular to ignorance. Each occurrence of a surprising event may change the state of ignorance. The relationship between the state of ignorance and surprise may be compared to the relationship between a stock and its flows. While the flows in terms of the emergence of surprising events can be observed, the stock in terms of ignorance is itself unknown. From this it follows that ignorance and surprise cannot be understood independently of each other.

In a formal mathematical framework, this relationship between surprise and ignorance has been recognized in information theory. There we may talk of the "surprisal" associated with new information, in a closely defined sense (Shannon and Weaver 1949; Proops 1987).[6]

We now offer a brief taxonomy of ignorance. (For a fuller taxonomy and extensive discussion of ignorance, see Faber, Manstetten, and Proops 1990.) Before we discuss this taxonomy, however, consider Figure 1.

We are often unaware of our ignorance and therefore feel no need for learning or research. We call this "closed ignorance." Closed ignorance may spring from unawareness of unexpected events or from false knowledge or false judgments. As long a person remains in a state of closed ignorance, its state cannot be recognized; only if some event forces the experience of surprise can one experience, ex post, the previous state of closed ignorance. Yet very often individuals (politicians, scientists, social groups) suppress as far as possible the possibility of surprise and therefore remain unaware of their state of closed ignorance. Closed ignorance, particularly in the form of pretended knowledge, is a great barrier to human cognition and insight, as well as to the solution of ecological and economic problems.

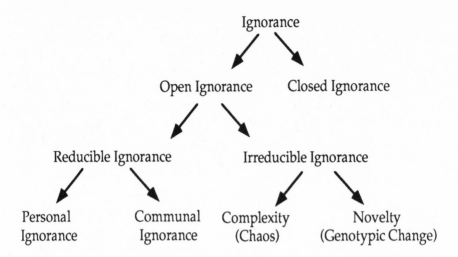

FIGURE 1. *A Brief Taxonomy of Ignorance*

If individuals (groups, societies) become aware of their closed ignorance (forced by drastic events or guided by a changed attitude), they reach a state of "open ignorance." In this state we will become attentive, for example, to events we had neglected earlier. Only in a state of open ignorance are we able to experience surprise to its full extent. Of course, we will try to understand surprising events by learning and research. But we are not only aware that we may generate new surprises by research and learning, but we know that we remain, in spite of our knowledge, essentially in a state of ignorance.

We must distinguish here between ignorance that is in principle "reducible" and ignorance that is in principle "irreducible." By reducible ignorance we mean that which is amenable to study, learning, and the application of scientific method. Irreducible ignorance is not amenable to these tools. We further subdivide reducible ignorance into "personal" and "communal" ignorance. We also divide irreducible ignorance into "ignorance from complexity" and "ignorance from novelty." We turn first to the two categories of reducible ignorance:

• Surprise because of *personal* ignorance means that an event is unexpected only because of an individual's state of knowledge. Knowledge that would eliminate this surprise is available within the society. Thus, by learning, personal ignorance may be over-come.

• Surprise because of *communal* ignorance means that an event is unexpected because there is no information available to a society concerning this event. By research, however, it would be possible to obtain this information.

We now turn to irreducible ignorance:

• Surprise because of ignorance through *novelty* occurs if the "potentialities" of a system alter. That is, the structure of the system underlying the event is changed without there being knowledge of a law governing this change. An example is the mutation of a gene or the invention of a new technique of production.
• Surprise because of ignorance through *complexity* means that an event is unexpected because, owing to the complexity of the underlying process, the information available is not (and will never be) sufficient to predict this event—perhaps because of insurmountable limits of computability. Systems that exhibit "chaos'"(Gleick 1987) have deterministic dynamics, for example, but are so infinitely sensitive to their initial conditions that they are impossible to predict in detail with finite computational techniques.

Ignorance through complexity can be attributed to systems that are computable, and thus predictable, in principle but not in practice. Ignorance through novelty can be attributed to changes in the structure of an objective system, where these changes are not predictable.

From this it follows that in a state of open ignorance—that is, without the barriers of closure—we are able to experience surprise. This enables an individual or a society to reduce, by learning or research, the state of personal or communal ignorance. Of course, learning and research may also lead to the discovery of new areas of ignorance. A case in point is the interaction between economic activity and the processes in the natural environment. This interaction lays bare our ignorance and exhibits surprise to a great extent, because only few people would have expected that our technological and economic activities, measures to improve our welfare and security, would lead to such environmental degradation. A concrete illustration is the use of chemical fertilizers and pesticides, which endanger the soil, groundwater, and, via food chains, also consumers, future generations, animals, and plants. We conclude that we need awareness of our ignorance and openness to surprise not only in our everyday life but also in our science and environmental policy. Given this perspective we now turn to some important implications for science.

The Concept of Evolution

When we are interested in obtaining a notion of an "open future," it is expedient to take recourse to the notion of evolution, for this emphasizes the openness of a process—that is, it admits explicitly the possibility of surprise. The notion of evolution is used in various contexts. Thus different sciences have different concepts of evolution—we speak of geological, biological, social, political, and economic evolution, for example, as well as the evolution of consciousness. Each science uses different concepts of evolution according to its different aims of study. Thus in biology the key concepts for evolution are mutation, heredity, and selection, while system theory uses the concepts of self-evolution, self-sustained development, and self-reference.

Evolution may be understood in a narrow and in a broad sense. The *narrow* sense presupposes experience and knowledge about the whole time sequence of certain processes in the past—the growing up of a young person into an adult, for example, or a seed into a tree. These evolutionary processes may safely be supposed to occur again and again in the future. Evolution in a *broad* sense is a process of change that starts either in the past, in the present, or in the future. Its characteristic is that it is open-ended insofar as: (1) we do not know if it will end, (2) in the case that it does end, how it will end, and (3) if it does not end, what kind of changes will occur. We recognize evolution to include explicitly processes of open-ended change. In our view, the key characteristics of evolution are process, change, and time. Thus we define: Evolution is the process of the changing of something over time. (See Faber and Proops 1990:chap. 2.)

The dichotomy between evolution in a narrow and a broad sense is artificial. For in reality there exists a continuous range of evolutionary processes between these two extremes: while processes of narrow evolution are in principle predictable and those

of broad evolution are unpredictable, in reality we, as partly ignorant human beings, are confronted with processes whose traits are
partly predictable and partly unpredictable.

Evolutionary Processes: Phenotype and Genotype

Evolutionary processes cannot be conceived of without an observer. Since human observers live in time—that is, in past, present, and future—they can only experience processes in these
subjective time categories.[7] There exist many processes, however,
where the explicit recognition of subjective time is not of much
importance. Therefore, processes can be classified into those
whose description is independent of, and those whose description
is dependent on, the past, present, and future. To the first class
belongs the fall of a stone. To the second class belongs the history
of Europe or the development of modern art. It is obvious that a
description of this history or this development depends on the period of time in which it is made. An essential distinction between
the two types of processes is that the former can be sufficiently
explained as realizations of a certain law, while the latter cannot
be explained by any known law.

A law, whether a natural or an economic law, is usually considered to be time-independent. Therefore, once the law is known the
processes governed by it are always predictable. They are identical
ex ante and ex post. Because processes like the historical ones,
however, cannot be explained completely by any law, they are not
predictable. Only their past parts can be described or narrated
with some certainty. Their future course can be neither explained
nor narrated: they are intrinsically open to surprise.

How can predictable and unpredictable processes be distinguished operationally? We suggest that processes are predictable
when their structure is time-invariant and known. This invariant
structure is the basis for predictability. Predictable processes can
be conceived of as a realization of an invariant and known structure. The fall of a stone, for instance, is a realization of the Law of
Gravitation. The growth of a tree is the realization of its seed, that
is, of its genotypic structure. The formation of the price of pota-
toes in a free market is the realization of the Law of Demand and
Supply. The innovation of a technique is the realization of its invention.[8]

But if the structure of certain processes changes and we do not know an invariant metastructure that governs the change, there no longer exists a basis for predictability. Such processes are intrinsically time-dependent and thus change in an unknown way. Hence processes that do not have an invariant and known structure are unpredictable. For example, the seed of a tree can be changed by mutation. The mutation and the corresponding process of its realization have the characteristic of novelty; therefore the occurrence of a mutation is unpredictable. From this it follows that the evolution of a new species of tree is, in principle, unpredictable. Similarly, the breakthrough invention of a new technique is the structural change of the set of technology and, again, is unpredictable. All the consequences of the invention of the steam engine were unpredictable, for example, at least up to the time it was invented.

In accordance with the terminology of biology, we distinguish two aspects of each process (Faber and Proops 1990, 1991). The underlying structure of a process is called its genotype; the realization of this structure is called its phenotype. Although the phenotype is the realization of its genotype, it is not causally determined solely by it, since there exist additional influences—for example, the environment for the growth of a tree. From this it follows that, even with given genotypes, often the phenotype can only be predicted in broad terms. Moreover, the meaning of genotype and phenotype changes according to the science in which they are applied. A genotype in biology, such as the genetic potential of a seed, is conceptually very different from a genotype in an economy, such as that expressed in a technique, an institution, or a consumer's preference ordering. The interplay of genotypes and phenotypes of different areas is of great importance for economic and ecological development.

Although all processes with genotypic changes are in principle unpredictable, it does not follow that processes with no genotypic changes (that is, with invariant structures) can be predicted. For instance, the weather cannot be predicted for more than a very short time period, although the genotype in terms of natural laws is known and constant (that is, time-independent). This is so not only because we are ignorant of genotypic changes but because many processes are so complex that human beings simply cannot predict them, as we know from chaos theory (Lorenz 1963).[9]

Some Applications: Resource Depletion vs. Pollution

The two main areas of concern in economy/environment interactions are the depletion of natural resources and the production of pollution. Here we wish to examine how our evolutionary framework, especially our concepts of novelty and ignorance, can be used to analyze these two areas.

Let us consider a nonrenewable resource. If we know the total quantity that appears to be extractable (that is, the stock) and how much of it is exploited each year (the flow), we can predict the time when this resource will be exhausted. We note that the magnitude of the flow depends also on the states of individual and social ignorance. Thus, for example, oil had almost the character of a free good before the first oil crisis in 1973, because its price was so low. The more individuals and societies at large became aware of the finite nature of oil, the less oil was used per capita. The main means to increase this awareness was the sharp rise in the price of oil. On the basis of the knowledge of the stock, the flow, and the influence of the price, it is possible, at least in principle, to conceive political and economic measures to manage the extraction of a certain resource such that it is reasonable.

In our considerations so far we have regarded only individual and social ignorance that leads to a waste of resources. We have seen that the price mechanism is an effective means to reduce this species of ignorance. While these kinds of ignorance are detrimental, there are other kinds that can be related to beneficial novelty. New deposits of a resource may be discovered, new recycling methods may be invented, new production methods using less of the resource may be found, and completely new technologies may be invented using material that heretofore has not been considered to be a resource at all. The emergence of such kinds of beneficial novelty will be furthered by the means of the price mechanism, for as the resource is used up its price will rise. This increase in price will generate a search for further reserves of the resource (prospecting), a search for techniques using less of the resource (resource-saving invention), and a search for techniques using cheaper resources (which may eventually lead to resource-substituting invention).

Besides these price-induced inventions, there will also be considerable non-price-induced research, which will tend to move the economy further away from its natural constraints. While in the short run many of these developments may be predictable, it is exceptionally difficult to predict them in the long run. This is well illustrated by the history of industrialization. Looking only at the use of resources, then, we may conclude that the problems arising from ignorance have at least a chance of being solved by the activities of humankind.

Let us now consider pollution. Ignorance is of much greater importance here than for the problem of resource depletion. Looking at water, air, soil, and waste we can observe a similar pattern of occurrence. At the beginning of a certain economic activity we often are unaware that pollution is produced. When the pollution is first recognized, we do not consider it to be really harmful. If it is recognized to be detrimental, we often do not know why it is detrimental, as is the case with forest death. From this it follows that we do not know what we can do. Finally, if we really know what we can do, it may be too late.

Here the surprise that emerges is in terms of new relationships between economic activity and the natural environment. This surprise is potentially, and usually, deleterious. Examples include the greenhouse effect, the hole in the ozone layer, algal bloom from fertilizer runoff, acid rain and its effects on lakes. All of these are relatively new phenomena, and all were unexpected and unwelcome.

Countering the effects of pollution by "internalizing" them via the price mechanism (for example, by leveling charges) would certainly help. Unlike market-led resource depletion, however, while the scarcity of the environmental resource, such as clean water or air, would appear in its planning valuation ex ante, the effect of a new pollutant is initially unknown because of ignorance. Therefore, the "price" of the pollutant, and thus the corresponding cost of using it, would only be knowable ex post. Thus resource use can be said to generate scarcity, which is reflected in a market price, which in turn is likely to generate beneficial novelty. On the other hand, new pollutants are themselves a source of deleterious novelty and generate only slowly, and often not at all, a search for a system of market pricing to encourage the reduction of their emission.

In summary, then, there exists an inherent asymmetry in the recognition of problems of natural resource depletion and the degradation of the environment; this is so because of different forms of ignorance. Hence an attitude that is appropriate for the treatment of resource problems may be completely inappropriate for pollution problems.

Relevance for Ecological Development

Pollution problems require an attitude of openness. One prerequisite for a new attitude of openness is this: we must see clearly the prevailing conditions of our present situation and accept them. We have to recognize the Faustian traits of modern societies. This holds also to a great extent for ecologists. They, too, have tended to cope with environmental dangers within the area of reducible ignorance. For example, water treatment plants have been built and cars have been supplemented with catalysts. In our terminology these measures spring from a Faustian attitude. It is important for all of us to recognize how far we are involved in the Faustian world and ourselves express a Faustian attitude.

Although we consider certain Faustian measures absolutely necessary, we do not believe that they are sufficient. For we cannot expect to solve our environmental problems if we try to tackle them in the same Faustian spirit of pretended omniscience and omnipotence that humankind has used during industrialization. A Faustian attitude cannot cope with irreducible ignorance. The complexity created in society and nature by industrialized economies is so encompassing that we cannot expect that science and technology will help us to offer more than partial solutions. We must accept that humankind has not only been unable, up to now, to control many essential natural and social processes according to its needs, but will never be able to control all essential processes.

From this it follows that we must recognize our ignorance and accept it. This is difficult because we are often so deeply involved in Faustian projects and Faustian measures that we are unable to recognize our true ignorance. Moreover, it is almost impossible to conceive one's true ignorance in an adequate manner. For even if we are aware of it and are open to the corresponding surprises, it is impossible to talk about them as they really are. For as soon as we try to speak about our ignorance and explain it, we must have a conception of it in terms of knowledge, however slight it may be, and thus our openness is, so to speak, overcast by a shadow. Hence there is an intrinsic difficulty in drawing concrete conclusions—especially with regard to environmental and economic policy—from the awareness of ignorance. This crucial awareness of ignorance and the consequent attitude of openness constitute the main conclusion of our essay.

As a consequence, the emergence of novelty and the admission of ignorance should not be an "add-on" to policymaking but rather, its cornerstone and central tenet. The case for such a stance is overwhelming, we think, in considerations of economy/environment interactions in the long run. This approach leads us to the following conclusions:

• A simple belief in technical, economic, and social progress stands in contrast to an attitude of openness. Although progress is possible, we cannot rely on progress to solve all our environmental problems. Hence our plans should be open for progress but should not depend solely upon it.

• Any plan should be open to revision whenever novelty occurs. Planning should be a continuing process taking into account ignorance and emerging novelty—not only in practice, but also in models of planning and decision making.

• We should value, and plan for, flexibility in response to currently unforeseen possibilities. Planning for systems to operate near the limits of their capabilities may be optimal in the short run, for example, but it severely limits flexibility of response for the long run. (See the seminal paper by Koopmans 1964; see also Weizsäcker and Weizsäcker 1984.)

CONCLUSION

We are suggesting a new approach to environmental policy in which economic activities are conceived as embedded in open ecological systems and in which these systems develop and maintain their own self-organization. Therefore economic activities are acceptable only to the extent that they do not destroy the health—the capacity for self-organizing activity—of the ecological system within which the economy operates.

We have argued that, because of the different types of novelty and ignorance they present to us, pollution problems may prove more recalcitrant than resource availability problems. Resource problems can be foreseen and incentives can be arranged to encourage appropriate, ameliorating, economic responses; pollution problems, however, may attack the health of the system itself and limit human opportunities for innovative responses to change. What is called for in this new approach is not more and more futile attempts to control natural systems but more and more wisdom in terms of accepting our radical ignorance. This would allow us to discriminate against human impacts that will limit opportunities for present and future generations. A reorientation of environmental policy—away from attempts to bring more and more of nature within the human economic sphere, where it can be controlled, toward an open view of the future through the acceptance of our ignorance—is a first step in creating a more rational approach to environmental policy.

Acknowledgments

We are grateful to Jens Faber, Helmut Lang, Armin Schmutzler, Ramon Margalef, and Robert Ulanowicz for their comments. We are particularly grateful to Bryan Norton for his advice and extensive comments on the final version, and for his thorough, perceptive, and generally excellent editing.

References

Aristotle. 1984. *The Politics*. Trans. C. Lord. Chicago: University of Chicago Press.

Augustine. 1961. *Confessions*. Trans. R. S. Pine-Coffin. Middlesex: Penguin Books.

Binswanger, H. C., M. Faber, and R. Manstetten. 1990. "The Dilemma of Modern Man and Nature: An Exploration of the Faustian Imperative." *Ecological Economics* 2:197–223.

Ehrlich, P. 1989. "The Limits to Substitution: Meta-resource Depletion and a New Economic-Ecological Paradigm." *Ecological Economics* 1:9–16.

Faber, M., and R. Manstetten. 1988. "Der Ursprung der Ökonomie als Bestimmung und Begrenzung ihrer Erkenntnisperspektive." *Schweizerische Zeitschrift für Volkswirtschaft und Statistik* 2: 97–121.

Faber, M., and J.L.R. Proops. 1985. "Interdisciplinary Research Between Economists and Physical Scientists: Retrospect and Prospect. *Kyklos* 38:599–616.

———. 1990. *Evolution, Time, Production and the Environment*. Heidelberg: Springer.

———. 1991. "Evolution in Biology, Physics and Economics: A Conceptual Analysis." In S. Metcalfe and P. Saviotti (eds.), *Evolutionary Theories of Economic and Technological Change*. London: Harwood.

Faber, M., R. Manstetten, and J.L.R. Proops. 1990. "Humankind and the World: An Anatomy of Surprise and Ignorance." Discussion paper. Department of Economics, University of Heidelberg and University of Keele.

Faber, M., H. Niemes, and G. Stephan. 1987. *Entropy, Environment and Resources: An Essay in Physico-Economics*. Heidelberg: Springer.

Faber, M., J.L.R. Proops, M. Ruth, and P. Michaelis. 1990. "Economy-Environment Interactions in the Long Run: A Neo-Austrian Approach." *Ecological Economics* 2: 27–55.

Gleick, J. 1987. *Chaos: Making a New Science*. Middlesex: Penguin Books.

Goethe, J. W. 1908. *Faust, Parts I and II* . Trans. A. G. Latham. London: Dent & Sons.

Jaynes, E. T. 1957. "Information Theory and Statistical Mechanics." *Physical Review* 106:620–630; 108:171–190.

Keynes, J. M. 1963. "Economic Possibilities for Our Grandchildren." In J. M. Keynes, *Essays in Persuasion*. New York: Norton.

Koopmans, T. C. 1964. "On Flexibility of Future Preferences." In M. W. Schelly II and G. L. Bryan (eds.), *Human Judgements and Optimality*. New York: Wiley.

Kubat, L., and J. Zeman (eds.). 1975. *Entropy and Information*. Amsterdam: Elsevier.

Lorenz, E. N. 1963. "Deterministic Non-Period Flows." *Journal of Atmospheric Sciences* 20:130–141.

Lovelock, J.E.L. 1979. *Gaia: A New Look at Life on Earth*. Oxford: Oxford University Press.

Plato. 1953. "Timaios." In *The Dialogues of Plato*, vol. III. Trans. B. Jowett. Oxford: Clarendon Press.

Prigogine, I., and I. Stengers. 1984. *Order Out of Chaos*. London: Heinemann.

Proops, J.L.R. 1987. "Entropy, Information and Confusion in the Social Sciences." *Journal of Interdisciplinary Economics* 1:224–242.

———. 1989. "Ecological Economics: Rationale and Problem Areas." *Ecological Economics* 1:59–76.

Shannon, C. E., and W. Weaver. 1949. *The Mathematical Theory of Communication*. Urbana: University of Illinois Press.

Solow, R. M. 1973. "Is the End of the World at Hand?" *Challenge* 10:39–50.
Weizsäcker, C. von, and E. U. von Weizsäcker. 1984. "Fehlerfreundlichkeit." In K. Kornwachs (ed.), *Offenheit, Zeitlichkeit, Komplexität: Zur Theorie der offenen Systeme.* Frankfurt am Main: Campus Verlag.
Wilhelm, R. 1984. "Einleitung." In *Laotse: Tao te king.* Köln: Diederichs.

Notes

1. Thus the sinologist Wilhelm (1984:24–26) observes that the Chinese idea of "way" (Tao, the ultimate principle of the universe in Taoism) has many aspects in common with the idea of logos.
2. Of course, one may ask: Are there really two houses, a house of humankind and a house of nature, or is there only one house? And further: If there is only one house, is it the house of humankind (economy) or the house of nature (ecology)? Or is it a third one (Proops 1989:72)? We need not deal further with this metaphysical question here.
3. It is noteworthy that Faust, trying to translate the first verse of St. John's gospel, "In the beginning was the Logos," translates it in the following way: "In the beginning was the Deed" (*Faust*, Part I, Study). The course of the drama leaves no doubt that Faust sees his creation of the Faustian world as the true "beginning," so that the "Deed" is nothing but the Faustian deed.
4. That this anxiety has spread over many parts of society demonstrates that this far-reaching consensus has been increasingly lost. This has been shown by the demonstrations at Gorleben and Wackersdorf, for example, and the subsequent partial retreat of the nuclear industry in the Federal Republic of Germany.
5. In Greek philosophy, the notion of prediction was reserved for eternal matters (such as the celestial bodies). The prediction of future events on earth was a task not for philosophers, but for oracles.
6. There is a close, but often contentious, relationship between information theory and the thermodynamic entropy concept. One might view entropy as reflecting our state of ignorance about the world (Jaynes 1957; Kubat and Zeman 1975).
7. One has to distinguish between subjective and objective time categories. The latter consist of all time categories that describe a time sequence—for instance, earlier or later—as well as all relations between durations. In contrast, the former consist of the categories past, present, and future. While the meaning of objective time

categories remains always constant, the meaning of subjective time categories changes with the moment of time in which the observer is living.

8. One must be aware that the two notions (realization and potentiality) have different conceptual statuses in our examples, according to the different area in which they are employed. In spite of this, we have purposely used them in order to develop a common approach to different areas of knowledge (see Faber and Proops 1990:35).

9. Here we have analyzed two of the three elements of our definition of evolution—process and change. The other element, time, is extensively analyzed in Faber and Proops (1990:pt. III).

5

Environmental Existentialism

TALBOT PAGE

This chapter considers the question: how do we value ecosystems? My motivation is practical, not philosophical. There are many decisions that need to be made concerning habitat preservation, biodiversity, human populations, and the like, and we need concepts of valuation for such practical decisions. There are, of course, many concepts of valuation, but I will argue that we should do more than simply choose one and apply it. I begin with a metaphor comparing perspectives on valuation to villages of ideas. My purpose is to challenge the comfortable idea that the right approach to the valuation problem is to be found in one's own discipline (too bad for the guys in the other disciplines.) I then outline several traditional approaches—from institutional processes (economic, political, and legal), from philosophy, from biology and ecology, and from health. These approaches have homes in various academic disciplines, and if my metaphor is successful the concepts outside your own discipline will not be rejected out of hand quite so quickly. In this section I list the basic characteristics of the various approaches as they relate to the problem of valuing ecological systems and explain in what ways they seem mismatched to that problem.

One of my main points here is that traditional approaches to valuation have an entity orientation. The objects of valuation, or the valuers, or both, are highly individuated. But the entity orientation, which is pervasive, is mismatched in valuing ecological systems because these systems not only have blurred boundaries but also shape the valuers: us. Of the traditional approaches I consider, one of them, an approach from concepts of health, appears to be less tied to an entity orientation and consequently less mismatched to the problem of valuing ecological systems. Although there are dangers of vagueness in an approach from health, the direction seems sufficiently promising. Thus in the following section I consider one more approach, an approach from a concept of meaning. Even more than an approach from a concept of health, the approach from meaning appears to avoid difficulties of entity-orientation and some of the dichotomies that cause trouble in valuing ecological systems.

Of the dichotomies, the most important is the one between valuation and decision. In this dichotomy it is often argued that one style of thinking should be applied to both valuation and decision. (The argument is often tacit and even the dichotomy is often sub rosa.) But in the approach from meaning, valuation and decision require two different styles, and both must be integrated in the practical management of ecological systems. At the end of section IV, I suggest that the approach from meaning strengthens the case for two-tier systems of managing ecological systems. In the first tier broad instruments (such as taxes or policies toward education of women) are chosen; in the second tier decisions are made concerning specific ecological systems (such as preserving a tract of land or banning ivory imports). The key point is that different considerations guide decisions in each tier. Since I discuss two-tier approaches elsewhere (Page 1977:145–213; Page 1991:58–74), I am brief here.

The main points of the chapter, summarized in the conclusion, are these: Traditional approaches to valuing ecological systems are mismatched to the ecological systems themselves; compared with traditional approaches, an approach from meaning suffers less from the mismatch; this approach collapses some of the dichotomies that appear to cause trouble in valuing ecological systems and offers at least some guidance for practical decision making.

VILLAGES OF DISCIPLINARY IDEAS

How we think about the question of valuation depends a great deal on the direction from which we approach it and the baggage we carry with us. When I lived in a village in Ethiopia, the people were hospitable and caring, but they warned me that the people in the next village were liars and thieves. When I visited the dangerous village, however, I found the people hospitable and caring, and they warned me to watch out for the liars and thieves of the first village.

As I sketch some of the alternative approaches to valuing ecological systems, I hope I demonstrate that there are differences in the villages of ideas and give an impression of what some of these differences are. But I will not do enough, I am sure, to get beyond a certain culture-boundedness. Just as it is natural to think that the other village's ways are untrustworthy, so too is it natural to think that the valuation ideas from another discipline are nonintuitive, ad hoc, not compelling. This bias often leads to superficial dismissal.

I suspect that the only complete way to overcome culture-boundedness in ideas is to "forget" one's own disciplinary heritage, become an intellectual child again, and grow up in the neighboring discipline's village for a while. There is a shortcut that gets us partway there: while you think that your own ideas are right and natural and the ideas from the neighboring discipline are shifty and without sensible foundation, it helps to realize the inhabitant of the neighboring village thinks just as passionately— but in reverse. It is freeing, for an interdisciplinary discussion, to speculate at the beginning that had we grown up in different villages our sense of the obvious rightness of our views, and wrongheadedness of others, would likely be quite different.

I am not arguing here for a pure cultural relativism. Our ecological problems are real, and some ideas about valuation and decision are more useful than others. At the same time I believe that our disciplinary culture-boundedness sometimes gets in the way. Valuing ecological systems is a new problem, one that has emerged because of the accelerated massive destruction of ecological systems. What we need at this time is the ability to stand free from our disciplinary heritages so that we can pick certain ideas and develop new ones and stitch together a new approach to valuing ecological systems. With this cautionary tale in mind, we are ready to consider several of the traditional approaches.

THE TRADITIONAL APPROACHES

In this section I outline several traditional approaches—from institutional processes (economic, political, and legal), from philosophy, from biology and ecology, and from health. These approaches have homes in various academic disciplines. I attempt to identify the basic characteristics of the various approaches as they relate to the problem of valuing ecological systems and I attempt to explain in what ways they seem mismatched to that problem of valuation.

From Institutional Processes

In making social decisions in this country, we rely on three basic institutional processes: markets or marketlike processes; legislatures and voting processes; and courts and legal processes. Each has its own culture and approach to problems of valuation. These approaches were not developed for valuing ecological systems but for distinctly different types of decision and management problems; nonetheless, when faced with the challenge of valuing ecological systems, people often bring to bear their familiar concepts. And it is fairly predictable that in valuing ecological systems economists will suggest economic concepts, political scientists political concepts, and legal scholars legal concepts.

Of the three approaches, the economic has received the most attention from environmentalists, and not cordial attention. Aldo Leopold describes his version of the economic approach in his A–B cleavage. Those in group A regard land (and other parts of nature) as soil (and other resources) to be used instrumentally for human ends. Leopold writes, with patent distaste, that "group A is quite content to grow trees like cabbages." In contrast, group B "worries on biotic as well as economic grounds about the loss of species like chestnut, and the threatened loss of white pine. Group B feels the stirrings of an ecological conscience" (Leopold 1949:259). To Leopold, "let the market decide" is the lodestone toward disaster.

Mark Sagoff agrees. He finds efficiency to be the central conception of value in the economic approach. To him, this concept of valuation leads to Pizza Huts, landfills, commercial strips, and the wasteland of modern living. He suggests that this is what we can expect from market-oriented consumers (Sagoff 1981a:229). But to Sagoff there is a higher order of beings, called citizens, who participate in political discourse and make better decisions for the benefit of the whole community. Sagoff believes that people in the role of citizens are more likely to be protective of the environment (Sagoff 1988:99–123).

Bryan Norton contrasts Leopold, who has an ecological conscience, with Gifford Pinchot, who was willing to grow trees like cabbages. According to Norton, Pinchot was a political realist as well as an economic realist; he was willing to trade off grazing rights in the national forests to save them from privatization and development. But while Pinchot participated in the political process and was instrumental in establishing 175 million acres of national forests, Norton finds three fundamental flaws in his economic approach. First, Pinchot emphasized the vital importance of raw materials and neglected "recreational, aesthetic, spiritual, and other less tangible values in his rush to maximize timber and grassland production. His version of utilitarianism was too materialistic, too narrow, and too inflexible." Second, Pinchot viewed resource production in an atomized, mechanistic way (growing trees like cabbages) and ignored complex ecological interactions. Third, Pinchot viewed his economic management as a value-free science and thus, presumably, immune to criticism (Norton 1991:4–7).

Unsurprisingly, John Hicks, Nicholas Kaldor, Ezra J. Mishan, and other economists who developed cost-benefit analysis tend to have a different view as to its core idea. (Cost-benefit analysis is the primary application of the economic approach often suggested by economists for valuing ecosystems.) As we shall see, in the view of the economists themselves, the core idea is remarkably similar to that in voting. To illustrate the idea, let me sketch it in terms of an actual decision faced by the World Bank. In one of its development projects, the bank is considering loans for building roads and cutting timber in a tract of land in Indonesia. If the project is undertaken, incomes in Indonesia will go up through tax revenue, employment, and exports of timber products. But rain forest habitat will be lost and quite possibly a particular species of butterfly along with other as yet unidentified species. It is easy to say, "Save that tract of land, save the rain forest." Rain forests are a rallying cry, almost a holy cause, these days. But decision makers at the bank will want to know on what grounds we are to save the rain forests. They will say, "We agree that rain forests are valuable, but how valuable? The practical question is this: how hard should we work to save this tract of land, how much should we sacrifice to maintain it? For if we preserve this tract, we give up the timber and resulting income we could have gained from it."

The economic approach (I mean here cost-benefit analysis) offers a way of addressing the question of how much. You are asked to compare two worlds. The first is the status quo: the world the way it is now. The second is identical with the status quo except for the change brought about by the project. In the comparison, you take into account the ramifications of the project, differences in income to you and others, differences in habitat, and so on; but except for the changes brought on by the project, the two worlds are the same.

Suppose that you value the first world more highly than the second.[1] (You value the status quo more highly than the world with the project.) Then you are asked what is the minimum you need to be compensated so that you would value the change (with the compensation) just as much as the status quo. If you value the world with the project more than the status quo, then you are asked how big a payment you could make in the changed world (with the project) so that you would just value equally the status quo and the world with the project (and its ramifications but less the payment). We ask these questions to everyone affected by the decision, which in principle could be everyone in the world. The economic criterion says that if the sum of all the compensations (to those who would lose by the project) is less than the sum of the equilibrating payments (from the gainers from the project), then the change from the status quo is worth making. The compensations measure the costs to the losers; the equilibrating payments measure the benefits to the gainers.

This is a bare bones account of the economic approach. I will fill it out a little by considering some of the most common objections to it. First, it is often argued that the economic approach is excessively narrow and mean-spirited because it assumes that people's valuations are only self-regarding and thus the approach does not count a person's altruistic valuation of others' well-being.

The claim is a misunderstanding of the approach, but it is easy to see how the confusion can arise. It often arises from economists' explanations of the approach, and indeed I have just provided an example in my rain-forest scenario, a sketch that is fairly standard among economists.

Suppose, in the example, that I live in Indonesia and I personally am benefited by the project (my income goes up). I believe the rain forest will be damaged very little, but I also believe that the project will further impoverish many people who are already poor. Placing value on a more equal distribution of income, on balance I value the status quo more highly than the change. Now you ask me how much compensation do I need so that I would value the change with the compensation just as much as the status quo. I answer "no amount." For each dollar of compensation I get, the income distribution becomes worse and the project with the compensation becomes worse, not better, compared with the status quo. It is easy to conclude that for the compensation story to make sense there must be a tacit assumption that only self-regarding valuations count, not valuations concerning income distribution and the well-being of others.

But instead of agreeing with this narrow interpretation of the economic approach, let me reword it a little more abstractly and accurately. Again we compare the two worlds: the status quo and the project with all its ramifications. But now we ask if there is some redistribution in the changed situation that would make everyone value the changed situation (with the redistribution) more highly than the status quo. If the answer is yes, then the project is worth doing, even if the redistribution is not made. This general compensation test (called the criterion of potential Pareto improvement) permits people to have altruistic valuations of others' well-being.

Second, it is often objected that because the compensations are potential rather than actual, the economic approach may lead to bad distributional consequences. Most economists would agree that this is a serious objection. As a result they would suggest that other criteria besides the compensation test be used for important policy decisions, especially other criteria concerning distributional justice.

Third, valuation is income-constrained, and poor people have tighter constraints than rich people. Poor people might settle for the timber products from Indonesia and value species preservation quite modestly (in income compensation terms) because they are poor and a small amount of compensation makes a big difference to them. Thus even if there is actual compensation, there can be bad distributional consequences from the economic approach— and, as a result, bad decisions. This objection is similar to the second and the response is often the same: other criteria should be used in addition to the compensation test when there are important distributional consequences. The objection is distinct from the second, however, because it arises even when there is compensation.

Fourth, people often find it hard (or even impossible) to compare two different worlds and decide which is better. There may be incommensurabilities in trying to compare the extinction of a species with increases in employment. The usual response is that the problem of incommensurability is less when there are small changes or low risks. (For example, you might readily compare the risk of dying from jaywalking with the saving of ten seconds of time and value the jaywalking more highly.) But in valuing ecological systems, the changes may be large or the probabilities of serious harm may be large, and thus incommensurability may be a serious problem for the economic approach.

Fifth, where social decisions are made by the World Bank or regulatory agencies following the economic approach, the aggregation of values is done by proxy and estimation. Environmentalists and others are often suspicious of the experts who take on this task of estimation. What prevents these experts from following their own tastes and ideologies?

Sixth, there is the problem of counting future generations' valuations. Typically, though not always, in the economic approach only those in the present generation have their valuations counted and aggregated. The well-being of those in the future is taken into account by how much we in the present altruistically value their well-being. This last objection is distributional at heart and thus is related to the second and third objections.

These objections are not usually directed at voting processes, at least not with the same intensity. (Some form of voting is the primary application of the political or legislative approach to valuation and social decision.) But there is a striking similarity between the economic approach and voting as a process of valuation. To see the similarity with voting, consider a direct democracy such as a New England town meeting and a decision whether or not to acquire and protect a tract of land. Each eligible voter in the town is asked to compare two worlds. The first, the status quo, is the world without the town's decision to acquire and protect the tract. The second, the change, is the world with the project—the decision to acquire and protect the tract. Suppose that you are an eligible voter and you value the first world more highly than the second. (You believe that the tract costs too much and there is little environmental value in the land.) Then you vote against the motion to acquire and protect the tract. If you value the second world more highly than the first, you vote for the motion. The voting criterion, in the form of majority rule, says that if more people value the second world more highly than the first (and vote for it), then the change from the status quo is worth making. If the voting process is indirect, as in the U.S. Congress, the individual Congress members make the same comparison between the two worlds and vote according to their individual valuations. Of course, the comparisons of value may include the desires to represent constituents, to increase the probability of reelection, and so on.

To social choice theorists, the essential difference between valuation through the economic approach and valuation through voting is that in the economic approach it matters how much more an individual values one world compared with another while in voting it only matters which world is valued more. Intensities of valuation matter in the economic approach, not in voting. When there is symmetry in valuation intensity (one person's strong valuation of the first world is balanced by another's strong valuation of the second), the two conceptions of social valuation will lead to the same decision. The similarity in the concepts of valuation between the economic and voting approaches suggests that the foregoing objections to the economic approach may apply to the latter approach as well, although some differences may arise from the essential difference relating to valuation intensities.

The first objection we dismiss: both approaches as concepts of valuation allow self-regarding, other-regarding, and even ideal-regarding individual valuations. Sagoff and others have argued that discussion during a political process tends to bring out other-regarding valuations, whereas market interactions tend to bring out self-regarding valuations. Perhaps this is true. But the economic approach applied to valuing ecological systems takes place outside markets—for example, in conference rooms of the Agency for International Development, the World Bank, and regulatory agencies—and in these forums there is also discussion, argument, and consensus building. On the other side, those who have been to a New England town meeting will recognize that sometimes the most narrow self-interests can arise in that most direct of democracies. In other voting contexts, the very process of discussion sometimes seems to nurture polarization and self-regarding valuation.

The second objection—concerning the possibility of bad distributional consequences—applies to the voting approach as well, and indeed there is an enormous literature on the "tyranny of the majority."

The third—concerning income constraints—would seem not to apply to the voting approach and its absence would seem to be a principal virtue of the approach. It is often pointed out that a poor person's vote counts just as much as a rich person's. But it is often overlooked that the income constraints can affect voting valuations too. Poor people might value the project more highly than the status quo because they are poor and need the extra income from the project, and for this reason they might vote for the project. In fact, at least indirectly, this may be the most common path toward the destruction of ecological systems, especially in Third World countries.

The fourth—incommensurability—also applies to the voting approach, in which people also compare two whole worlds. It may apply with a little less force to the voting approach however, because "more" is easier to compare than "how much."

The fifth—lack of accountability—applies less directly in institutions relying on voting. Legislators are accountable to the public, whereas the economists in the Office of Management and Budget (OMB) do their work accountable to the public only very indirectly through the president (and experts in the World Bank still more indirectly).

The sixth—the problem of future generations—applies to both economic and voting approaches in the same way.[2]

I hope I have convinced you that the two approaches, economic and voting, are surprisingly similar even with their real differences. Both approaches start with individual valuations and then aggregate them in straightforward ways. Both are holistic in the sense that whole ecosystems and their ramifications are being compared (whole alternative worlds) and not just parts of an ecosystem. Both are anthropocentric in the sense that humans are doing the valuing and it is human values that count.

A major concern among environmentalists is to make the valuation concepts less human-oriented—to get other things besides humans and human interests to count. To a legal scholar such as Christopher Stone, a natural way of doing this is to extend the concept of a person. In law, persons are not limited to humans—for example, corporations are legal persons too. Stone suggests that we treat certain environmental entities as legal persons, as well, giving them legal rights and guardians to argue in their behalf in courts (Stone 1974). When I first read *Should Trees Have Standing?*, this idea seemed intriguing and almost bizarre, but part of the surprise was that he cited examples where the approach has actually been implemented.

The approach makes the most sense where the environmental entities can be well defined. The motivating case for *Should Trees Have Standing?* was Mineral King Valley, which became a defined entity through a court case and a proposed development project. But where the entity is less obviously a tract of land—for example, in the Tellico Dam case—should each snail darter be considered a legal person, the collective species of snail darter, or its habitat? It would seem most sensible to focus on the habitat or on the ecosystem contained by overlapping habitats. But one of the lessons of biology is that habitats and ecological systems are not entities. One system merges and overlaps with another without clear boundaries. If there were many "environmental legal persons," there would presumably be overlapping claims of rights and many invitations to litigation.

One wonders how close Stone's conception of valuation is to that of the economic and voting approaches. Of the three, the legal approach is the most process-oriented. The idea is that if we set up the right legal process, we will accept the outcome. But Ronald Coase's "The Problem of Social Cost" (Coase 1960) suggests a basic commonality between economic and legal processes. In addressing pollution problems, Coase suggests giving "property" rights for clean air or, alternatively, giving rights to pollute—he argues that for economic efficiency it does not matter which way the rights are defined as long as they are defined. Define the rights and let them bargain, he says. Coase contends that human interests, in the form of economic production or environmental quality, will be expressed in the bargaining setting backed up with potential legal recourse.

Stone says something a little different: give the trees rights and guardians, and let the guardians litigate with the developers. For Coase the rights go to humans; for Stone the rights go to trees. But in the analogy with corporations, only figuratively do the rights go to the "legal fiction" (corporation); they actually go to the officers of the corporation and their human beneficiaries. Perhaps in an analogous way interests of individual humans are being furthered by creating rights for environmental entities; in this sense, Stone's view like Coase's is similar to the economic and political approaches. It is not so clear whether trees have interests—if they do, Stone's view is more genuinely different from the other two institutional approaches. It is less of a leap to believe that snail darters have interests and are harmed when their habitat is destroyed. But again the question is: how do we value these interests and harms? Are "interests" and "harms" the sensible concepts for the problem of valuing ecological systems?

From Philosophy

In analyzing concepts of valuation and trying to develop a value theory that is less human-oriented than the institutionally based approaches, environmental philosophers have often followed the traditions of moral philosophy. There are two main traditions: one based on duty and going back to Kant, and the other utilitarian and going back to Mill, Bentham, and Hume. It is helpful to think of the first as focusing on the value of *motivation* of action and the second as focusing on value of the *consequences* of action.

Peter Singer is an example of the latter group: a utilitarian in the consequentialist tradition (Singer 1979). At least for many animals he answers affirmatively the question "Are harms and interests sensible concepts of valuation?" As long as an animal is sentient (has the capacity to suffer), it has interests (not to suffer), and it may have other interests and desires as well. In this way Singer extends the concept of interests from humans outward to a larger circle and in this sense his approach to valuation is less human-oriented than the economic and political concepts. His system of valuation rests on the normative idea that there should be "an equal consideration of interests." A mouse, being less sentient than a chimpanzee, will have less strong interests, of course, so even on a balance scale of "equal consideration of interests" the mouse may come out the worse. Under Singer's approach, it is arguable that trees are not sentient, and thus have no interests, and thus are protected only to the degree that they instrumentally further the interests of sentient creatures. Stone would be more generous. He argues that a lawn suffers harm when it is neglected and turns brown in the summer. But we may ask with Singer if this is not just an anthropomorphic attribution. We would suffer if we were desiccated and died in scorching summer sun, but perhaps the grass, having no neurological capacity to feel pain, feels none.

For many environmental philosophers, Singer's circle of protection is neither wide enough nor strong enough. To many, the system is too biased toward larger central nervous systems. This may be so, but here I am worried by another problem. Singer's extension of utilitarianism from humans to sentient creatures is still entity-oriented. We are thinking about valuing ecological systems, and it is not so clear how we can use an "adding up" criterion based on countable individual entities. In other words, in Singer we have the problem of needing countable entities, each one constituting a source of value (or utility or interest). The problem of needing entities to get the system going is the same as for the three institutional approaches. Ecological systems are not easily definable either as individual entities or as parts or boundaries.

In moral philosophy there has been a marked shift from utilitarian perspectives toward views depending on rights and duties—the Kantian tradition (Hart 1979). Here one of the key distinctions is between price and dignity. Things that have price are things that can be replaced by others as equivalent. But some things are "beyond all price" and have no equivalent. These are the things that have dignity (Kant 1959:53). For Kant, rational beings capable of following the moral law are beyond price and have dignity. Such things deserve respect and are "ends in themselves." Just as Singer extends the utilitarian idea from humans to sentient creatures, environmental philosophers such as Paul Taylor have extended the Kantian idea of dignity from rational beings to life forms, mountains, and Gaia herself. In doing so, Taylor shifts the locus of moral worth from the capacity for rationality to "teleological centers of life" (Taylor 1981:172–178).

This move, as such steps are called in philosophy, has the effect of putting environmental objects beyond the reach of cost-benefit analysis and the economic approach. According to this neo-Kantian view, it is not sensible to ask people to trade off an Indonesian tract of rain forest for an exchangeable good such as money or to follow the economic approach of comparing two whole worlds, one with the rain forest and one without but with higher incomes and compensations or payments. Viewing a natural thing as having dignity (putting it beyond price and marketlike tradeoffs) seems much the same idea as viewing it as having "intrinsic value," in the way deep ecologists do. Both perspectives, it seems to me, implicitly rely on a subject/object dichotomy. Subjects are those things that have dignity or intrinsic value (and those things we do not trade off in markets or marketlike procedures). Objects are things we can use instrumentally for our own benefit.

In viewing many natural objects as having intrinsic value, we of course do not solve the problem of valuation or decision. Are we to protect every natural object having intrinsic value? What happens when protecting one object with intrinsic value means harming another? The usual answer is that while we have duties to protect things of intrinsic value (and dignity), our duties are not all equally compelling. Our negative duties not to harm are more compelling than our positive duties to give aid to something of intrinsic value. Our perfect duties are more compelling than our imperfect duties.[3]

Economists will recognize here the problem of opportunity cost. By doing one action, performing one duty, another is foreclosed. By comparing two whole worlds, one with the project and the other without, changes in ecological systems are compared holistically. All the ramifications are valued together. In contrast, by rejecting the aggregation of values within each of the whole worlds, the neo-Kantian and the deep ecologist approach the decision problem piecemeal, intrinsically valued entity by intrinsically valued entity, and duty by duty. The task of deciding which duties to follow (and how far) is made more difficult because the neo-Kantians often try to avoid looking at the consequences of their alternative actions. (Kant thought that judging actions on the basis of consequences corrupts the purity of the motivations.)

It is an interesting question whether the Kantian tradition is inherently less holistic in valuation than the economic approach, but that is not my main concern here. Again, I am concerned with the entity orientation of the Kantian tradition. In that tradition individual things have dignity (or intrinsic value). Only individuals have duties. And as in the utilitarian tradition, of which the economic approach is one branch, this entity orientation makes it difficult to value ecological systems.

From Science

We take a long step away from institutional processes and moral philosophy when we approach the valuation question from the perspective of science, primarily ecology and biology. Here one gets the impression that the scientists start out at a disadvantage because of their training and acculturation. After years of focusing on tasks of description, explanation, and prediction, after years of admonitions to be objective and value-free, ecologists, biologists, and engineers are now asking how to value ecological systems. This is the same handicap faced by economists who were persuaded by Lionel Robbins and Milton Friedman that the task of economists is to describe, explain, and predict, to be value-free and objective and scientific.[4] To some extent this is the heritage of the logical positivists, who believed there could be clean distinctions between fact and value.

The self-identification of economics as a positive science left cost-benefit analysis and policy analysis in a curious position, since both are frankly normative. The compromise was that the criterion of Pareto optimality, while a normative idea, was accepted because it was so "universally acceptable" it hardly qualified as a value claim at all. Other normative ideas would be treated as by neutral technicians—if some policymaker asked an economist how to achieve, say, a more equal distribution of income, the economist as technician would analyze the ways of getting it. But, according to Robbins and Friedman, it is not for the economist to argue for normative ideas (other than Pareto optimality, which is the same as efficiency in this context).

The situation is one step worse for the scientists who have traditionally cast themselves outside the role of value discourse entirely. (At least the economists have a toehold in their concern with cost-benefit analysis.) Given this heritage, the scientists have done what might be expected of them. They have taken certain descriptive features of ecological systems and then given these features a value interpretation. For example, a measure of energy accumulation is defined. Then, the argument goes, an ecological system that sequesters more energy than another, under the definition, is better than the one that sequesters less. Or alternatively, some descriptive measure of complexity is defined. Then it is argued that one ecological system, which is more complex than another, is better than the other.

The procedure is one of anointing descriptive parts with value. There is a certain ad hoc quality. Complexity, per se, leaves me cold. Why not start with a concept that seems to carry something of value more directly—for example, sustainability? We could then say that complexity is often to be protected because it often promotes sustainability. But then it is not just any old sustainability that seems important, some kinds seem more valuable than others. Which kinds? Why? Valuable only to humans? We are off to a discussion of value for which the biologists, ecologists, and engineers have had little training. In response to the need to evaluate and protect ecological systems, scientists are increasingly drawing on other disciplines for ideas on valuation. In doing so, scientists have certain advantages because they are often closer to the applied problems and materials than anyone else.

From a Concept of Health

In the Environmental Science and Engineering program at UCLA, students start at the masters level and earn a doctorate. The doctorate does not require a dissertation, and the graduates are not intended to go into academic research and teaching careers. Instead, as the name suggests, the graduates are intended to become practitioners, doctors to heal the environment.

Valuing ecological systems in terms of a concept of health is an appealing idea, and the intuitions of some scientists lead toward it. We think of health as a harmonious relationship among the parts of the body and between the body and the outside world. Focusing on the relations among things takes us away from the subject/object dichotomy that causes so many problems in valuing ecological systems. The debate between intrinsic value only for humans or intrinsic value for humans and other environmental entities looks less urgent when neither entity has intrinsic value but value emerges through the proper interaction among things. Moreover, relations among things, rather than the things in isolation, seem to be the proper focus of attention when describing ecological systems. The concept of homeostasis, taught in medical schools as a normative concept, has a direct parallel with the notion of stability in ecological systems.[5]

How do we tell when one ecological system is more healthy than another? We do have certain signs indicating when a woods is thriving, when a beehive is "strong," or when an ecological system is stable. Such indicators may boil down to the indicators of energy mobilization and complexity. But we, as a society, have focused most of our attention on human health. When we move from a concept of human health to ecological health, two problems arise. First, individuals go to doctors, announce that they are not well, tell of symptoms, experience pain directly, and once treated tell when they are feeling better. Complaints of course make it easier for the doctor. As ecological systems are mute, in healing the environment the practitioner must be exceptionally skilled in reading signs. Perhaps veterinarian is a metaphor more accurate than doctor.

Second, and more fundamentally, a concept of a human's (or an animal's) health is embedded within a larger environment. Because we are adapted to a larger environment that has remained roughly the same over evolutionary time, our arms move and they move over specific ranges. When we have a limited range of movement, it is easy to recognize this loss of function as outside the norm, label it unhealthy, and suspect a torn rotator cuff. The outside environment may vary day-to-day, but over evolutionary time our response to it is so patterned that "healthy norms" are readily definable. What makes this concept of health tractable is the relative constancy of the outside environment to which we have adapted. The outside environment becomes a benchmark.

But how do we compare one outside environment with another or one whole ecological system with another? There are no constant reference points from which functions and norms can be defined. How does one take the temperature of Gaia herself (and relative to what)? And if one could do so, how does one interpret it in a value sense? I am not saying that this second problem is an insuperable one for constructing valuation concepts for ecological systems—only that deriving valuation concepts based on notions of health is likely to be different and more challenging for ecological systems than for individual humans.

AN APPROACH FROM MEANING

It would seem that "meaning" is an even more elusive concept than health. Yet the two are related and a concept of meaning has useful properties for the problem of valuing ecological systems. Like health, meaning is a relational concept. A sense of meaning is a prerequisite to valuation and meaning is a collapser of dichotomies. The first dichotomy to go is the one between subject and object. As a relational concept, meanings arise from what is going on between our inside world and our outside world. There is no clean boundary between these two worlds in this interaction. This shift of focus from the isolated parts to the overlapping interactions will sound familiar to ecologists and others who study ecological systems.

When thinking about a concept of meaning, the centuries-old dispute among moral philosophers as to which is the right foundational normative idea—duty or consequence—seems a bit misplaced. I suspect that an existentialist like Camus would say there should be no either/or choice between good intentions and good actions. We create meaning through action and once we have a sense of meaning, actions flow naturally in an "authentic" way. We find ourselves in a meaningless world and we must create our own meanings through actions for which we are the authors.

When I think of an existentialist concept of meaning, I think of a Parisian resistance fighter in the early 1940s. The situation was much in doubt. Quite possibly any action one could take would be ineffectual; "authorship" of resistance acts was highly dangerous. Yet for a resistance fighter to sit out the war was meaningless. To take action brought life to one's own life, even if it brought physical death as well. I am reminded of a cancer patient whose future was very much in doubt. He looked at the abyss and gave others small gestures of connection and humor, snatching life from death. And in a less visibly dramatic way, I am reminded of certain environmentalists. To them too the future is in doubt. Their own actions may be ineffectual; the environmental movement is no place for those who need assurance that the consequences of their actions will "make a difference."

When looking at the abyss, we choose actions (or medical treatments) for the best likelihood of success, even if that likelihood may seem small. Actions are chosen in spite of the doubt; they are chosen because of the doubt; the very doubt gives them a deeper meaning. There are, of course, many who look at the evidence of environmental problems and do not see an abyss. Perhaps they are right, the problems minor, and strong actions unwarranted. When there is so much uncertainty, there is room for genuine doubt. There are others, too, who look at the evidence of environmental problems with a sense of paralyzing hopelessness.

Irvin Yalom, an existentialist psychotherapist, says there are two common reactions to deep threats to one's life: denial and fear (Yalom 1989). Both reactions can cloud intention and preclude action. I am not concluding that Julian Simon's optimism is a reaction of denial from severe environmental problems; instead I am saying that denial is a common reaction to a view of the abyss. More than that, I am saying that in an approach from a concept of meaning, the problem of valuation is more difficult than in the traditional institutional approaches. In the institutional approaches, an individual person's valuations are taken pretty much for granted as exogenous and usable in the valuation of ecological systems. In an approach from *meaning*, individual valuations are taken as a starting point, inviting further inquiry as to their formation. A sign that we are on the right track is when individual valuations arise from an integration of intention and action, action and consequence. This hint of, course, is not enough to develop a concept of value, but it points in a direction and warns us from the opposite extremes of disconnection from fear and denial.

Another dichotomy to be collapsed is that between the positive and the normative—between the task of description, prediction, and explanation and the task of evaluation. In terms of a concept of meaning, the two tasks go together. Without a sense of meaning, there is no sense of relative importance, no priority guiding the positive tasks; even more debilitating, there is no reason to do the positive tasks. (As Hume put it, reason is the slave to the passions, and should be.) Going the other way, to achieve a sense of meaning one needs to participate in the world. And undertaking the positive tasks of predicting, describing, and explaining is a way of doing so.

To someone not brought up in the village of moral philosophy, it is curious to see in books on moral argument the typical split between arguments for a particular value system in the first part of the book and the dangling last chapter on "compliance" or "why be moral?" Here I think the institutionally oriented approaches have it better. It makes little sense to talk about value in isolation from institutional processes and incentives to implement these valuations. Further, in an approach from meaning, the dichotomy between market participant (consumer) and voter (citizen) is blurred. A person's values are shaped by his or her whole experience and only partially by the particular decision procedure, voting or marketlike mechanism, the person is participating in at the moment. In summary, an approach from meaning suggests the following collapses of dichotomy. For subject and object, neither concept stands in isolation; for consequence or duty, both are taken together; for normative and positive, both are taken interactively; for valuation and implementation, no separation; for consumer and citizen, a blurred mixture.

In an approach from meaning, the larger situation of value and decision is like that of an hourglass. The grains of sand at the top are in a complex spatial relation with each other and not linearly ordered. But when the sand moves through the narrow middle waist, it becomes ordered in a ranked linear fashion. The logic of value is like the logic of meaning. It is complex, interactive, not simply ordered. But the logic of decision is ordered. The word itself means to "cut off," "to kill," as in homicide, suicide, or pesticide. Making a decision to do one thing kills another alternative.[6] In their essence, decisions are rankings of things. Somehow the complex unranked relations of value and meaning come to the narrow waist of decision and ordered choice.

The idea of ranking is commonplace in economic and voting decisions. It may seem less applicable to legal decisions. But note that a jury may decide in favor of a plaintiff over a defendant because the plaintiff's version of the facts was "more likely than not," or the case was "closer" to one precedent than another, or there were "more compelling" social policy reasons. These are words of comparison and ranking, and the final result is a decision. The process is often painful. And in a mundane way I am reminded of the process of writing. My original thoughts are not linearly ordered, but complexly patterned; they keep shifting and I have trouble seeing how they are related and which is most important. But then, when the paper is written, there it is in a linear order, one word in front of another. And if the paper is any good, the reader can tell what I thought was important and sense the whole pattern from the linear ordering.

Our problem, then, is to go from the richly patterned top part of the hourglass to the narrow waist. There is one last dichotomy to go. The choice is not between either the logic of value, where the relations are richly patterned, or the logic of decision, where relations are simple rankings. The problem is to put the two together. The challenge is not to choose between the top part or the middle part of the hourglass, but to construct the whole hourglass. If this approach from meaning is sensible, this problem of putting the two parts together arises in all types of policymaking. But in managing ecological systems, this challenge is especially difficult because the systems are so complex and vital for our well-being.

So far we have a diagnosis. What are the recommended policy directions? It seems clear that we cannot make decisions on ecological systems one-by-one in isolation. If we do so they will gradually be nibbled away. And the decision context is not just one ecological system in comparison with others. The decision context should include the institutional processes and incentive systems too. The battle cry "Save this rain forest, no matter what" will be overwhelmed by sheer human numbers if the population is not stabilized.

Earlier I suggested that basing a concept of ecological system evaluation on physical complexity is not as useful as basing it on a notion of sustainability. I said this because sustainability is closer to something we value directly. But then I left vague how "sustainability" should be defined. We can be a little more specific now. It is hard to think of a stronger prerequisite to a meaningful sense of life than a sustainable future connected to both our cultural and our biological heritage. In talking about these two heritages we are talking on general terms, and thus we are seeking general characterizations. Yet the anthropological literature is replete with examples of how a society can become severely disconnected from its cultural (and economic) heritage—with accounts of the Ik, the Bushmen, the Native Americans. In a general way all these stories are the same.

In a parallel way scientists are developing accounts of how we are becoming disconnected from our biological heritage. Again the accounts need to be specific case studies but generalizable. Where the disconnections are most threatening we must concentrate our protective actions. As we do these things, generalized policy instruments, so neglected by environmentalists, may come more sharply into focus. These include, for example, energy taxes (to promote conservation), education of women (to delay family formation and promote population stabilization), and income leveling (to make economic policy actions more feasible and equitable). Such actions will not save ecological systems in the short run, but they will help in valuing and managing them in the long run.

The approach from meaning invites a direction of inquiry toward the problem of valuation, one that puts the small and large together. With this approach, when you look at the small decision (say the particular decision on a particular tract of land) you see the large (the long-term decision context of human population). When Thoreau walked past a clear New England brook, he looked at the sand at the bottom and saw the stars above.

CONCLUSION

Valuing ecological systems is a hard problem partly because it is, roughly speaking, a new problem. A hundred years ago, even fifty years ago, people did not think much about ecological systems, still less about valuing and managing them. When people did think about ecological systems, they seemed large, resilient, and capable of taking care of themselves. Now that the world population and economy have become large enough to transform ecological systems, we recognize that our ability to value them is an urgent issue. Without being able to value, compare, and choose, we cannot manage, preserve, and take care.

Each of the approaches I have sketched has, I think, something to contribute in developing ways of valuing ecological systems and each has weaknesses. I offer here the barest of sketches but hope that it points to some of the core ideas. The biologists have a connected view of ecological systems, but not a value theory to go with it. The moral philosophers have value theories, but they seem too entity-oriented (individuated) for valuing ecological systems. The institutional approaches have the virtue of integrating valuation and implementation and they are decision-oriented, but they are entity-oriented. And especially the economic approach takes individual valuations as given or exogenous. The approach from health is relational and has normative concepts built in, but the concept must be developed more and integrated more with the other approaches in order to become a more practical guide to making choices about ecological systems. The approach from meaning goes a step further as a relational concept and integrates things previously viewed as dichotomous. The problem of valuation differs from that of decision, just as the parts of an hourglass differ. Valuation begins with complex patterns; decision ends with linear rankings. For good decisions the two need to be brought together. How we bring them together depends on how we conceive the nature of our connection with ecological systems, how we derive our sense of meaning, and who we want to become.

In discussing the problem of what we owe to future generations, Brian Barry surveyed the traditions of political philosophy (Barry 1977). He found a constant among the differing approaches of Hobbes, Locke, and Rousseau—they are all based on notions of symmetrical power and reciprocity. He found these traditions inadequate for the problem of future generations, which have neither power nor reciprocity with us. One of the primary ways we treat future generations is through our management (or mismanagement) of ecological systems. Thus Barry's concern and ours are interlocked. And like Barry, I find something mismatched between the traditional approaches to valuation and the problem at hand: valuing ecological systems. The mismatch comes not only from the entity orientation of the traditional approaches but also from another common thread.

Each of the traditional approaches to valuation is a closed system in the sense that each rests on something taken to be solid and exogenous. I have already mentioned exogenously taken preferences (values) for the economic approach. Similarly rights, utility, biological function, adaptation to a fixed outside environment, complexity, and so on are taken to be solid starting points. But ecological systems are open in the sense that they are self-defining and evolving and shape our values. What we need for their valuation is a more open approach to valuation. To do this we can begin by taking values themselves as endogenous. They are shaped by our whole experience. Our most crucial choices concern the values we want to grow into, the environment we want to place ourselves in. These choices will define us, our children, our grandchildren. The choices are hard because they are open. We are choosing ourselves. By choosing what ecological systems to protect and in what ways, we not only express today's values but choose tomorrow's.[7] We are existential environmentalists.

This all may sound very grand, but it points in useful directions. I will close this chapter by identifying three. First, there is a parallel between valuing ecological systems and valuing educational and health systems. All three are infrastructural in the sense that they define our future. We do not establish school systems on the basis of existing preferences of six-year-olds. We ask instead what kind of people we want them to become. Similarly when we establish the health care system, we begin by asking what kind of society we want to become. Just by the amounts we spend on them, it is clear that we value health care and educational systems highly. But in comparing ecological systems to education and health, we will not find ready-made guidance from the latter two. We are having trouble valuing and managing health and educational systems and thus we are having trouble with all three. This should not be a surprise, since they are similar at the foundational level. But seeing how they are similar and borrowing what successes there are—conceptually and practically—may be helpful.

Second, a potential problem in valuing ecological, health, and educational systems is the problem of self-referencing, a problem often characteristic of open systems. (If today's experience shapes tomorrow's sense of meaning and value, and tomorrow's values shape decisions determining the next day's experience, how do we tell what are good values for today?) One way of dealing with self-referencing problems is to separate the key concepts into tiers.[8] In a constitutional democracy we separate the basic institutional rules, such as the constitution itself and the role of the Supreme Court, from a second tier of ordinary institutions, such as the state legislatures and markets. Markets and other ordinary, second-tier institutions function within the environment provided by the first-tier, constitutional institutions.[9] As part of the second tier, we need the decentralized processes that the economic and the other institutional approaches offer. We have just seen, in the Soviet Union and East Europe, how grave environmental harms can arise when the mundane institutions of markets and voting (and their attendant valuation approaches) are neglected. But we need not rely on markets and voting institutions automatically or in toto. We can embed the everyday decision processes and valuation approaches in a larger first-tier perspective, one that is shaped by our sense of meaning and shapes the second-tier institutions.

Third, the approach from meaning suggests a combination of long-run and short-run strategies. The first tier provides a roughly stable set of circumstances for the second tier to operate in. For ecological systems this means stabilizing human populations as well as energy and material flows. Without such long-run strategies, decisions on setting aside this tract of land or protecting that endangered species will be overwhelmed and pointless. The long-run strategy, which requires a decision horizon of centuries, must be complemented by a short-run strategy focusing on endangered species acts, preserves, and breeding programs. These can be seen as holding actions. (The Galapagos Islands are a key example.) If there is little biodiversity and little of our ancient heritage a century from now, the long-term strategy itself will be rather pointless.

Acknowledgments

I wish to thank Doug MacLean, Bryan Norton, Mark Sagoff, and the students in "Environmental Theory and Philosophy" for many helpful comments and discussions.

References

Barry, B. 1977. "Justice Between Generations." In P.M.S. Hacker and S. J. Raz (eds.), *Law, Morality, and Society: Essays in Honour of H.L.A. Hart.* Oxford: Clarendon Press.

Caldwell, B. 1982. *Beyond Positivism: Economic Methodology in the Twentieth Century.* London: Allen & Unwin.

Coase, R. 1960. "The Problem of Social Cost." *Journal of Law and Economics* October:1–44.

Hart, H.L.A. 1979. "Between Rights and Utility." *Columbia Law Review* 79:828–46.

Kant, I. 1959. *Foundations of the Metaphysics of Morals.* New York: Library of Liberal Arts Press.

Leopold, A. 1949. *A Sand County Almanac.* New York: Oxford University Press.

MacKinnon, B. 1986. "Pricing Human Life." *Science, Technology and Human Values* 11 (Spring):2.

Norton, B. G. 1992. "Epistemology and Environmental Values." *Monist* (forthcoming).

Page, T. 1977. *Conservation and Economic Efficiency.* Baltimore: Johns Hopkins University Press.

———. 1991. "Sustainability and Problems of Valuation." In R. Costanza (ed.), *Ecological Economics: The Science and Management of Sustainability.* New York: Columbia University Press.

Ramsey, F. 1950. "Truth and Probability." In R. B. Brathwaite (ed.), *The Foundations of Mathematics and Other Logical Essays.* New York: Humanities Press.

Sagoff, M. 1981a. "At the Shrine of Our Lady of Fatima, or Why Political Questions Are Not All Economic." *Arizona Law Review*: 23.

———. 1981b. "Economic Theory and Environmental Law." *Michigan Law Review* 79:1393–1419.

———. 1988. *The Economy of the Earth.* New York: Cambridge University Press.

Singer, P. 1979. *Practical Ethics.* Cambridge: Cambridge University Press.

Stone, C. 1974. *Should Trees Have Standing? Toward Legal Rights for Natural Objects.* Los Altos, Calif.: William Kaufmann.

Taylor, P. 1981. "The Ethics of Respect for Nature." *Environmental Ethics* 3 (Fall):197–218.

Yalom, I. 1989. *Love's Executioner.* New York: Harper Perennial.

Notes

1. Conventionally economists use the term "preferences" instead of "values," with the understanding that preferences can be ranked in terms of importance. Frank Ramsey, one of the first economists to analyze rankings systematicaly in terms of valuation and uncertainty, used the term "value" (Ramsey 1926:39). To noneconomists "preference" often has the connotation of "unimportant valuation." To suggest that we are talking about important valuations, and not "mere preferences," I will use Ramsey's wording.
2. How would a defender of the economic approach respond to these objections and comparisons? I suspect that someone like Mill (not exactly a cost-benefit analyst but a contributor to the economic approach) would respond that the differences between the economic and voting approaches, arising from the fundamental difference between "more" and "how much," boil down to problems of

distribution and accountability. (In other respects the two approaches share the objections.) If the economic approach is to be used more, relative to voting, in evaluating vital decisions concerning the environment, health, and education, then we should take a more egalitarian income distribution more seriously. With a more equal distribution of income comes a more equal distribution of "dollar votes" in a cost-benefit evaluation. The more basic and vital the decision making, the more concern with fairness in the distribution in decision-making power. It is not for nothing that Mill developed utilitarianism and the economic approach as instruments of reform. I suspect that Mill would also say that the cost-benefit analysts in OMB should be made more publicly accountable.

3. A positive duty is to take some action; a negative duty is to refrain from some action. For a discussion of positive and negative duties see MacKinnon (1986:32–36). A perfect duty is from one individual to another; an imperfect duty is from an individual but not to any specific individual. For example, one might have an imperfect duty not to harm the environment. For a discussion of perfect and imperfect duties see Sagoff (1981b). Both authors apply their analyses to public policy decisions.

4. For a discussion of the impact of positivism on economics see Caldwell (1982).

5. Moreover, Hippocrates' injunction "First, do no harm" has the same asymmetrical weight put on commissions, compared with omissions, as does the distinction between positive and negative duties.

6. Yalom (1989:14) makes this point.

7. Sagoff (1988:103) makes a similar point about the endogenicity of value in quoting Ulysses: "I am part of all that I have met." Sagoff says that values "express the kind of people they are and the sort of society they wish to be" (p. 117). In fact, one could interpret the central problem of his *Economy Of The Earth* as working out implications of the endogenicity of environmental values.

8. This is the traditional way of dealing with Russell's paradox, for example. Thus you do not allow sets to be defined in terms of being members of themselves.

9. For a discussion of a two-tier approach to the basic infractructure, see Page (1977: pt. III).

6

Environmental Therapeutic Nihilism

EUGENE C. HARGROVE

The notion of ecosystem health has developed on the model of human health as an aesthetic perspective. Everybody can tell if a person "looks" healthy, and usually that is what it means to "be" healthy. Nevertheless, doctors can frequently determine that a person who looks healthy really is not. Seen in this way, the ability to perceive health in a person depends largely on technical knowledge. Even when there is agreement about technical knowledge, however, the diagnosis is still frequently a matter of perception. It is for this reason that second opinions in medical matters are encouraged. A person who appears healthy to one doctor may appear to be seriously ill to another. On occasion, patients are treated for the wrong illness for many years or are treated for an illness when they are actually healthy. Despite the increase in technical knowledge, "looking" healthy remains an important guide that may encourage treatment or discourage it.

In "The Land Ethic" Aldo Leopold uses a notion of ecosystem or land health that is nontechnical, since it is not tied to a specific theory that defines ecosystem health. He gives us few technical clues for determining ecosystem health; rather, he produces three undefined values: integrity, stability, and beauty (Leopold 1949).

How does one use these values to make a specific determination of health? Leopold leaves this question unanswered, thereby making the determination dependent on the perception of the individual observer (who may or may not have relevant technical knowledge). This position would seem strange to anyone who was aware of Leopold's detailed discussions of ecosystem management in his previously published book, *Game Management*, where he lists the various factors in natural systems that can be manipulated in isolation and gives detailed advice on how to proceed (Leopold 1933). This incongruity can be traced to the fact that Leopold had lost faith in his factor management approach long before he came to write "The Land Ethic." Because Leopold embraced a conception of land health in the context of his disillusionment with the reductionist management approach, his position is analogous to a medical position developed in the nineteenth century called therapeutic nihilism. Although medical therapeutic nihilism has had a significant influence in many academic disciplines, there does not seem to be any direct influence on environmental management. Nevertheless, similar conditions—in medicine in the nineteenth century and in ecology in the twentieth century—seem to have produced parallel though unrelated results. These similarities are important because they may suggest a technical role for an otherwise nontechnical conception of ecosystem health in environmental management.

A MEDICAL PARADIGM

Therapeutic nihilism is the view that it is better to rely on the healing powers of nature, rather than trying to cure the patient, because efforts to cure the patient are likely to create complications that further reduce the patient's health.[1] Originally therapeutic nihilism was invoked as an experiment to determine whether folk medicine of various kinds had any effect. In accordance with this medical policy, traditional folk treatments were ignored even when no alternative scientific medical treatment was available and records were kept to see if patients could recover without them. Through these systematic experiments, it was possible to show that most patients recovered equally well with or without the application of traditional folk medicine. Consequently, folk medicine was ignored across the board.

Therapeutic nihilism as a form of medical practice achieved its grim classical form in the middle of the nineteenth century when doctors concluded that the limits of medical knowledge had been reached and that treatment of a wide range of illnesses actually reduced their patients' quality of life and, in fact, seriously endangered their lives. In Vienna, for example, the chances of surviving childbirth were inversely proportional to the level of care provided. Women frequently died in the maternity ward of the city hospital run by the doctors. The survival rate was better, but still unacceptable, in the ward run by midwives. If birth occurred on the way to the hospital, however, the chances of recovery were 100 percent. The widespread belief that treatment was bad for patients was supported by poor results in surgery and by illness brought on by poor hygienic conditions at a time when germ theory was not yet known. Believing that they could not cure their patients, doctors shifted their attention away from treatment and focused on diagnosis, studying diseases and other medical problems for their own sake. This change in focus dramatically altered the doctor/patient relationship. Ignoring the suffering of their patients, doctors withheld pain medicine because it distorted the symptoms of the medical conditions they wished to study.

THE PARADIGM APPLIED TO ECOLOGY

The notion of therapeutic nihilism found its way into professional environmental management when problems with the scientific management of natural systems arose in the 1930s and 1940s—for example, deer irruptions on the Kaibab plateau. In the National Park Service, therapeutic nihilism is now embodied in the policy called "natural regulation." Moreover, it is a belief commonly held by most environmentalists and the general public and is routinely invoked in nature preservation arguments. At the popular level, environmental therapeutic nihilism is known as Barry Commoner's third law of ecology: "Nature knows best" (Commoner 1971:41). According to this axiom, "any major man-made change in a natural system is likely to be detrimental to that system." This so-called law is widely and uncritically accepted by the general public and most environmentalists on the basis of a comparison between professional ecology and amateur watch repair. According to this analogy, attempts by ecologists to solve problems within natural systems are comparable to—and no more effective than—random thrusts into the works of a watch by a person with no knowledge of watch repair. In nature preservation arguments this law is presented, not as a belief, but rather as a factual claim—that ecologists do not have the knowledge, and will never have the knowledge, to manipulate natural systems without creating disruptions with unanticipated damage. Therapeutic nihilism permits environmentalists to avoid discussion of environmental values because it can be presented as a factual claim that policymakers tend to respond to more favorably than to value-based arguments. Values in this century have come to be viewed as subjective, arbitrary, irrational, emotional, and meaningless.

One drawback of this application of environmental therapeutic nihilism is that the theory's factual basis may be undermined as we learn more about the behavior of stressed ecosystems and develop new technical solutions to environmental problems. A dramatic breakthrough—a paradigm change—is probably not necessary. (Ecological knowledge has already improved enough without such a breakthrough that deer irruptions rarely occur today.)

There are other problems with this therapeutic nihilist argument. First, the belief in therapeutic nihilism as a *fact* has created a schism between environmentalists and theoretical ecologists, who believe that environmentalists are claiming that the limits of ecological knowledge have been reached, that ecological science will never improve, and that ecologists will never become better scientists. It is unfortunate, given that ecology is the foundation of environmentalism, that so many environmentalists feel compelled to defend nature preservation in terms of an argument that demeans and underestimates ecology. Second, it is quite likely that environmental therapeutic nihilism, like its medical counterpart, is responsible for widespread disregard of animal suffering in natural systems, as exemplified in the Bambi syndrome: accusing people of being pointlessly emotional (irrational) about the lives of wild animals. Presumably, this callousness arises from the belief that we are unable to relieve animal suffering in any systematic and large-scale manner without destroying natural systems—turning these systems into managed parks and the animals into domesticated animals—and as in medicine probably serves no useful purpose.

It might be supposed that—as ecological knowledge increases and economically feasible technical solutions to ecological problems become available and are employed routinely in the field—environmental therapeutic nihilism will be abandoned because it is based on a belief that is probably false. The history of medical therapeutic nihilism suggests otherwise, however. Although it is customary to ridicule it as an example of how medical science can become derailed, therapeutic nihilism was never actually abandoned and, in fact, has a methodological role today in twentieth-century medicine. If therapeutic nihilism were to evolve in a similar manner in ecology, it might then continue to have a significant influence on the concept of ecosystem health.

Medical researchers today, even though they know that the limits of medical knowledge have not been reached, still employ therapeutic nihilism selectively. In accordance with historical therapeutic nihilism, diagnosis is supposed to precede treatment—and treatment only occurs in the context of theoretical knowledge. When such knowledge is lacking, diagnosis takes precedence over treatment on the grounds that treatment will distort symptoms, thereby inhibiting the accumulation of knowledge about the medical condition and delaying the development of a cure. For example, patients are usually not given antibiotics until an appropriate germ has been discovered in a throat culture. In accordance with this approach, there was considerable pressure not to treat the victims of Legionnaire's disease until the specific cause of the disease was found. The original patients were treated, but only because they were initially misdiagnosed by general practitioners as having some standard form of pneumonia. The fact that many of them recovered with the use of antibiotics was not considered to be scientifically significant until a previously unidentified bacterium was finally found and connected with the disease. When knowledge of a particular medical condition becomes available that suggests specific treatment, medical researchers then write papers in medical journals arguing that therapeutic nihilism should no longer be applied.

It is hard to say whether such a conception of therapeutic nihilism will be adopted in ecology at both theoretical and practical levels. If it is, however, it will be explicitly coupled with a conception of ecosystem health that is nontechnical, given that therapeutic nihilism will be applied whenever knowledge and theory are inadequate. This conception of health will largely be a perceptual one, the kind discussed at the very beginning of this essay, since it will not be tied to any particular ecological theory. Nonetheless, this conception of health will not necessarily be counterintuitive or nonscientific. First, although assessments about ecosystem health will be based on value judgments as well as factual evidence, there will be room, as in medical practice, for conflicting expert opinions. In some cases, however, theoretical knowledge may make some judgments better than others. Second, the reliance on ecosystem health—for example, the reliance on the resilience of ecosystems when technical knowledge is lacking—will be part of a technical method for gaining new knowledge.

It might be argued that the analogy between medical and environmental therapeutic nihilism is inappropriate because the former depends on an organismic conception of health that is out of fashion, perhaps permanently, in ecological theory. While I agree that in this respect the analogy is inappropriate, this difference between health in ecology and medicine may not be very important. David Ehrenfeld, in Chapter 7 of this book, indicates that developing a conception of ecosystem health may be fraught with problems "because it is extremely difficult to determine a normal state for communities whose parameters are often in a condition of flux because of natural disturbance."

Actually, in terms of medical therapeutic nihilism, there are two conceptions of health—the normal state from which the patient has deviated and the resilience that may permit the patient to recover. When an ecological problem is clearly the result of an external intrusion of a foreign plant or animal, treatment can proceed without therapeutic nihilism being invoked, for in such cases diagnosis would appropriately precede treatment and the treatment would be in the context of appropriate knowledge and theory. Cases involving cyclic change, however, cannot be treated in this manner. The first question that will arise is whether the system has deviated significantly from normal conditions—the healthy state. The second question will be whether the system can survive the cyclic change without assistance—requiring a judgment about the resilience of the system to overcome the disturbance on its own. Ecologists might choose not to interfere because they judge the system healthy enough to recover on its own and because they want to study the natural effects on the system. If, however, the system is judged to be too unhealthy or too stressed to recover, they may choose to interfere, treating the symptoms without proper knowledge and theory, or, if treatment is judged to be too arbitrary or recovery too unlikely, invoke therapeutic nihilism and observe the downward crash of the system to its conclusion in the hope of learning something useful for the future. In all of these cases, ecosystem health is tied not only to a popular conception of health, but also to a scientific method of inquiry.

CONCLUSION

Although this conception of ecosystem health appears on the surface to be similar to Ehrenfeld's recommendation that ecosystem health be employed in such a way that it is not tied to a specific theory, it is an improvement upon that recommendation in some respects. First, by associating ecosystem health more with method than with theory, conflicts between the popular conception of ecosystem health and a scientific conception are eliminated. Viewed this way, technical conceptions of health are not replacements for the popular conception but, rather, enhancements or clarifications tied to specific ecological phenomena. Many technical definitions of ecosystem health for specific systems and subsystems could be developed that do not conflict with the popular conception or with each other. (In accordance with Wittgenstein's notion of family resemblance, they would not have any particular necessary or sufficient features [Wittgenstein 1958].) Second, following the medical model for therapeutic nihilism in which it has come to be applied to specific areas in medicine, rather than to medicine as a whole, it provides a way in which environmentalists can still invoke therapeutic nihilism selectively in their preservationist arguments—as a factual claim about the state of knowledge in a specific area in ecology—without having to employ the dubious claim that the limits of ecological knowledge in general have been reached.

The employment of therapeutic nihilism in this manner not only goes a long way toward mending the schism between environmentalists and ecologists. It also offers the possibility of a group of interrelated conceptions of ecosystem health for environmental policy that is flexible enough to take into account improvements in ecological knowledge while remaining consistent with the basic intuitions of the general public.

References

Commoner, Barry. 1971. *The Closing Circle.* New York: Knopf.

Hargrove, Eugene C. 1989. *Foundations of Environmental Ethics.* Englewood Cliffs: Prentice-Hall.

Leopold, Aldo. 1933. *Game Management.* New York: Scribner's.

————. 1949. "The Land Ethic." In *A Sand County Almanac.* New York: Oxford University Press.

Wittgenstein, Ludwig. 1958. *Philosophical Investigation* (3rd. ed). New York: Macmillan.

Notes

1. For a more detailed discussion of therapeutic nihilism, but without specific reference to its possible usefulness in ecology, see Hargrove (1989: chap. 5).

Part II

Science and Policy

7

Ecosystem Health
and Ecological Theories

DAVID EHRENFELD

The popular environmental movement often justifies its idea of ecosystem health by reference to an ecosystem theory that has been abandoned by most ecologists—the organismic theory. But if ecosystems, because of the shifting nature and number of their component species, do not resemble organisms, how can we speak of ecosystem health? Is health an appropriate concept for an entity whose composition varies from place to place and does not necessarily remain stable from one half-century to the next, even when free of significant human influence? Despite the problem posed by ecological theory, ecosystem health and illness seem to be palpably real and observable qualities. We risk damaging a useful, indeed necessary, idea by associating it too closely with a particular model of ecosystem structure and function. Ecological theories come and go; fads are as common in ecology as in other branches of science. To avoid giving the appearance of sanctioning ecosystem destruction by being unwilling to come to grips with the concept of health, ecologists should continue to use the term, defining it, when necessary, in a general way that is independent of particular theories of community or ecosystem function. As a working concept within ecology, stress is widely used because it is more definable and testable than health. Nevertheless, health should remain in the ecologist's vocabulary because of the important, nonquantifiable values it contains and the point of contact it establishes with nonscientists.

When I first became aware of environmental problems, in the 1960s, it was commonly said by the environmentalists of the day that Lake Erie was dead. This was a vivid image, even though it is difficult to imagine a lake as a corpse, and it did much to arouse interest in pollution and other legitimate environmental concerns. Of course the lake was not dead, although its best friends were having trouble recognizing it. Perhaps it would have been better to say that it was in ill health.

The idea of health, which covers a spectrum of conditions ranging from being well to being dead, is still applied to communities and ecosystems by nonscientists and scientists alike. The new field of restoration ecology uses this terminology often, for the concept of restoration is itself based on the idea of health. In the introduction to *Environmental Restoration* (1990), edited by John J. Berger, I found, for example, the following sentences: "In the Northeast, thousands of lakes are dead or near death, and forests, too, are sick and dying from acid rain and other pollution. For the first time in human history, masses of people now realize not only that we must stop abusing the earth, but that we also must restore it to ecological health."

ORGANISMIC THEORY

The image of a healthy environment is compelling, natural, and probably ancient, yet it also has origins in the ecological thought of this century. In the classic textbook *Principles of Animal Ecology*, by Allee, Emerson, Park, Park, and Schmidt, published in 1949, there is a celebrated table comparing seventeen supposedly shared properties of cells, multicellular organisms, and communities. The properties include anatomy, ontogeny, regeneration of parts, senescence and rejuvenescence, and dynamic equilibrium. This is the organismic theory of ecology—the idea that communities are structurally and functionally like organisms. It is most closely associated with the name of F. E. Clements, who was active during the early decades of the twentieth century.

Insofar as the organismic theory makes us compare communities to individual organisms, it is a normative interpretation of what communities are like. Your Aunt Margaret's parakeet is always a parakeet—it conforms more or less closely to some sort of platonic abstraction of what a parakeet should be: in its youth a lively, joyous parakeet, in middle age a calm, reflective parakeet, in old age a nasty senile parakeet, eventually a dead parakeet, but always a parakeet. You would be surprised if you looked in its cage and saw that it was turning into a lichen or a wombat. So it is with communities in the organismic view. They have a recognizable identity, and in the final stage of community embryology or succession, that identity becomes fixed and normative: a prairie, a beech–sugar maple forest, a desert. Because communities have fixed identities, because they are normative like organisms, we can easily apply the normative idea of health to them: if they are functionally and structurally similar to their abstract ideal, they are healthy; if they deviate significantly, they are sick.

If the organismic theory of communities were still dominant, if the idea that communities have a normative, equilibrium position, a balance point, were still widely accepted, then the idea of ecological health would pose few problems and there would be less need for a symposium on the subject. But concepts change in ecology, as in other fields of knowledge. Since the 1920s, but especially during the past two or three decades, the idea of the nature of communities and ecosystems has undergone revolutionary revision. No longer are communities considered normative.

NON-EQUILIBRIUM THEORY

The first hint of this change was provided by the controversy over diversity and stability, which arose in the mid-1960s and early 1970s with the discovery that the least diverse communities are usually the most resilient ones, partly because they are made up of organisms that are tolerant of a wide range of environmental conditions (Denslow 1985). Although still controversial in some of its theoretical and applied aspects, the idea that diversity and certain kinds of stability can be inversely related is accepted by most ecologists. The side-effects of this discovery, which included a lingering, if totally unnecessary, problem for conservationists, are not relevant here. What is relevant is that for the first time disturbance was beginning to be widely seen as an essential feature of the existence of many ecological communities. The history of the last fifteen years of ecological research and thought has been a steady extension of this role of natural disturbance in the life of communities and a corresponding erosion of the normative view of community organization and function.

Among the kinds of communities and ecosystems now known to be frequently influenced by disturbance are temperate deciduous forests, many coniferous forests including boreal and coastal plain forests, tropical evergreen forests, grasslands, and certain aquatic systems. Even deserts and tundra are affected by natural disturbance (Pickett and White 1985). Disturbances occur spatially and temporally; in the latter case, temporal disturbance exists on time scales measured in hours, days, months, years, and millennia, depending, respectively, on whether one is looking at tides, fires, virus cycles, or climatic change as agents of disturbance. Even without anthropogenic disturbance communities and ecosystems are in constant flux. How, then, do we define health, which depends on a definition of a normal state, and how do we separate, except in the most obvious cases, the effects of human and natural disturbances? As Pickett and White (1985) state in the summary chapter of *The Ecology of Natural Disturbance and Patch Dynamics:* "An essential paradox of wilderness conservation is that we seek to preserve what must change." Equally paradoxical, in some cases we must work to preserve disturbance, such as frequent fires, in order to preserve what we have come to think of as the essential character of certain ecosystems (Karr and Freemark 1985).

The discovery, by ecologists, of disturbance has caught the public eye. On 31 July 1990, the lead article in the science section of *The New York Times* was headlined "New Eye on Nature: The Real Constant Is Eternal Turmoil." The writer of this article, William K. Stevens, was alert to the environmental implications of disturbance theory—particularly to the discrediting of the organismic notion that there is a balance of nature, a healthy equilibrium state in communities that have not been unduly affected by human actions. Although the article was careful to point out that even disturbance ecologists admit to "a kind of equilibrium on some scales of time and space," the subheads conveyed no such hesitation: "A Difficulty—Posing a Question: What Is Natural?" and "Empty Theory—Observations Find No Neat Balance."

Before examining the implications of this paradigm shift for the concept of ecological health, I want to look at this change in ecological theory in a different way. Natural disturbance has always existed in communities, and ecologists have known about it for a very long time. But prior to the 1970s the prevailing view was to ignore it in favor of data and interpretations that supported the idea of homeostatic, organismic balance. Why? I think it was because these ecologists were themselves part of a human environment that instilled a strong sense of a normative community, a balance. In the United States, at the time when the organismic theory developed, there was a powerful sense of what the equilibrium position of this national community was. This sense of a national equilibrium was not fundamentally challenged by either the Great Depression or the Second World War, both of which disturbances were seen as sicknesses or departures from a normal condition and treated accordingly without overwhelming feelings of hesitation or uncertainty. It is not surprising that ecologists during this period saw nature as made up of definable, homeostatic communities, much like their perception of their own country.

Then came Vietnam, the military-industrial complex, the crisis of urban life, the clash of materialism and idealism, the status quo versus the women's movement and the struggle for racial equality, upheaval in the universities, speciesism versus animal rights, and the decline of communal morality and the apotheosis of the individual. Where was the balance point? Is it a coincidence that the individualistic theory of community structure and the changing view of stability and diversity began to be widely accepted around the time that the movie *The Graduate* was first shown? Is it a coincidence that mathematical ecologists had a brief fling with catastrophe theory in the late 1970s and that they are now—as elements of the old Soviet empire group and regroup—preoccupied with chaos? I do not think these are coincidences. For twenty years, the idea of a national norm, an equilibrium position, has been breaking up under the influence of repeated disturbances. Nearly fifty years ago, W. Clyde Allee could write professionally about the organismic theory and protocooperation within animal populations, while in his private life advocating world unification and peace. I do not know what his theoretical position would be if he were alive in these times.

It would be naive to assume that today's nonequilibrium theories of community and ecosystem structure are likely to be the last word on the subject. But to observe that ecological theory appears to be culturally dependent and changeable is not to invalidate the present form of that theory. The modern perspective raises serious problems with the concept of environmental health.

PROBLEMS WITH THE CONCEPT

First, as I noted earlier, it is extremely difficult to determine a normal state for communities whose parameters are often in a condition of flux because of natural disturbance. David Schindler (1987), in a careful review of aquatic ecosystem responses to anthropogenic stress, observes: "The fact is that we usually do not know the normal range for any variable, at least for any time period greater than a few years. Even if well-designed monitoring programs were magically emplaced tomorrow, it would be years before we could confidently distinguish between natural variation and low-level effects of perturbations on ecosystems."

True, there are mitigating factors. Anthropogenic disturbances can be orders of magnitude greater than natural ones (or faster, if we are referring to disturbances such as climatic change); this can help establish at least a comparative equilibrium for the purposes of defining health. Or disturbance and major fluctuations at one scale of observation—for example, the population scale—may not be matched by fluctuations at a different scale—for example, the scale that integrates many populations: the community. Moreover, Schindler notes that both paleoecology and the use of experimental ecosystem perturbations could help circumvent the problem of defining a healthy baseline in the face of normal variation. And as Koetsier et al. (1990) have recently pointed out, despite the prevailing fad for nonequilibrium theories in ecology, many communities may behave in an equilibrium fashion. "To totally displace Equilibrium Theory with Non-Equilibrium Theory may be premature and counter-productive," they state. As part of their critique they offer an interesting observation: "No ecological data will ever show a variable to achieve and maintain the *exact* reference state it occupied before a disturbance. How closely the variable must approximate its previous state is a matter of interpretation. The subconscious biases brought to the study by the researcher may influence the interpretation of the experimental results." Nevertheless, the problem of nonnormative, nonequilibrium systems has been raised and will not go away.

The second problem with the idea of ecosystem health is that ecosystems have many functions and processes, not all of them strongly related to each other. A determination of ecosystem health can be a function of which process you are looking at, which in turn is determined by your own values. The accident at Three Mile Island may have been made much more serious by control room operators who had no way of knowing which of the many kinds of reactor variables to pay most attention to during the emergency. Some remained normal; others did not (Perrow 1984). Of course the same issue could be raised about human health, the determination of which is highly dependent on which variables are observed, yet we do not discard the idea of health for that reason. But again, the problem exists.

Third, a general word such as "health" can end up with all kinds of narrowing qualifications and can lose some of its original meaning if we apply it too rigorously to examples of specific communities. Because communities vary greatly, with some occurring at the equilibrium end of the range, some at the nonequilibrium end, and others at all degrees in between, the unifying idea of health can vanish as we redefine it from case to case. Any attempt to imbue ecological health with the rigor and specificity that will allow us to use it as a scientific tool may well strip it of the intuitive, general meaning that is its chief value. We have an example of this in medicine, where a five-year remission from cancer, following strenuous treatment of the disease, may be equated with health by medical statisticians, but not necessarily by the patients whose definition of health embraces a far larger universe of experience.

CONCLUSION

One way around all three obstacles is to avoid using the word "health" in ecology. Many ecologists, like Schindler, prefer the much more definable and concrete term "stress." David Rapport et al. (1985), in a paper entitled "Ecosystem Behavior Under Stress," conclude that "ecosystems exhibit considerable uniformity in their general response to various stresses. A core set of ecosystem-level signs of distress can be identified." Regardless of whether one agrees, statements like this can be made and tested because stress is a definable and measurable working term in ecology.

Nevertheless, to avoid giving the appearance of sanctioning ecosystem destruction by being unwilling to come to grips with the concept of health, ecologists should not be afraid to use the term at least occasionally, even though it may have no place in their daily research. There is no need to define it in most cases. And when it is defined, it should be explained in a durable, general way that is as independent as possible of particular ecological theories of community or ecosystem function. Such definitions may involve, in addition to the usual considerations of loss (or gain) of components and changes in structure, the rates at which changes in ecosytems have occurred. They will always involve informed but unmeasurable human perceptions of what is happening to the ecosystem.

Health is an idea that transcends scientific definition. Although it contains values that are not amenable to scientific methods of exploration, they are no less important or necessary because of that. Health is a bridging concept connecting two worlds: it is not operational in science if you try to pin it down, yet it can be helpful in communicating with nonscientists. Equally important, if used with care in ecology, it can enrich scientific thought with the values and judgments that make science a valid human endeavor.

Acknowledgment

I thank Joan Ehrenfeld for many constructive suggestions in the preparation of this manuscript.

References

Allee, W. C., et al. 1949. *Principles of Animal Ecology.* Philadelphia: Saunders.

Berger, J. J. (ed.). 1990. *Environmental Restoration.* Washington, D. C.: Island Press.

Denslow, J. S. 1985. "Disturbance-Mediated Coexistence of Species." In S.T.A. Pickett and P. S. White (eds.), *The Ecology of Natural Disturbance and Patch Dynamics.* San Diego: Academic Press.

Karr, J. R., and K. E. Freemark. 1985. "Disturbance and Vertebrates: An Integrative Perspective." In S.T.A. Pickett and P. S. White (eds.), *The Ecology of Natural Disturbance and Patch Dynamics.* San Diego: Academic Press.

Koetsier, P., P. Dey, G. Mladenka, and J. Check. 1990. "Rejecting Equilibrium Theory—A Cautionary Note." *Bulletin of the Ecology Society of America* 71(4): 229–230.

Perrow, C. 1984. *Normal Accidents.* New York: Basic Books.

Pickett, S.T.A., and P. S. White. 1985. "Patch Dynamics: A Synthesis." In S.T.A. Pickett and P. S. White (eds.), *The Ecology of Natural Disturbance and Patch Dynamics.* San Diego: Academic Press.

Rapport, D. J., H. A. Regier, and T. C. Hutchinson. 1985. "Ecosystem Behavior Under Stress." *American Naturalist* 125(5):617–640.

Schindler, D. W. 1987. "Detecting Ecosystem Responses to Anthropogenic Stress." *Canadian Journal of Fisheries and Aquatic Sciences* 44:6–25.

Stevens, W. K. 1990. "New Eye on Nature: The Real Constant Is Eternal Turmoil." *New York Times,* 31 July, pp. C-1–C-2.

8

What Is Clinical Ecology?

DAVID J. RAPPORT

Over the past decade and a half, interest in diagnosing the causes of ecosystem breakdown and coming to grips with strategies for rehabilitation of damaged ecosystems has been steadily growing. (See Holling 1986; Rapport 1989a, 1989b; Schaeffer , Herricks, and Kerster, 1988; Schaeffer and Novak 1988.) A decade ago, papers on the topic "ecosystem medicine" sparked debate about the aptness of medical metaphors, while today international societies for the promotion of "ecosystem health" (such as the newly created Aquatic Ecosystem Health and Management Society) attest to a growing acceptance of the approach.

It is fitting, therefore, to take stock of the development of this fledgling field, which we may here term "clinical ecology." To do so, I elaborate on the concept of health from an ecosystem perspective and examine some promising approaches to gauging the health status of whole ecosystems. After contrasting curative approaches with preventive approaches in the treatment of ecosystems under stress, I suggest areas of clinical ecology that remain undeveloped and need more rigorous definition if we are to advance the practice of clinical ecology.

WHAT CONSTITUTES A HEALTHY ECOSYSTEM?

The short answer to this question is that a healthy ecosystem is whatever ecologists, environmentalists, and the public at large deem it to be. If this sounds irreverent, it is in fact no less committal than the medical fraternity which has by and large discarded "objective" standards in favor of considerations such as "life goals" of the individual (Porn 1984) or the concept of health as a "modus vivendi" enabling imperfect humanity to achieve a rewarding and not too painful existence (Dubos 1968). In the ecological realm, one generally confers the connotation of "health" to a state of nature (whether managed or pristine) that is characterized by systems integrity: that is, a healthy nature exhibits certain fundamental properties of self-organizing complex systems.

These properties, in one way or another, derive from the early work of Ludwig von Bertalanffy. Von Bertalanffy (1950) characterized the evolution of complex systems in terms of four major attributes: "progressive integration," "progressive differentiation," "progressive mechanization," and "progressive centralization." Steedman and Regier (1990) have shown the relevance of these concepts to the evolution of aquatic ecosystems. "Progressive integration," for example, entails the development of integrative linkages between different species of biota and between biota, habitat, and climate. "Progressive differentiation" refers to the often observed progressive specialization as systems evolve biotic diversity to take advantage of abilities to partition resources more finely and so forth.

Such criteria provide necessary but insufficient conditions for determining the health status of ecosystems. Ultimately, as in human medicine, such determinations hinge on human values. That is, what is "desired" or "healthy" must also take into account social and cultural as well as ecological values. These values may differ markedly among various segments of society. Native peoples, for example, value the integrity of the forest as a "cultural home," one that permits the survival of traditional ways of gathering food, spiritual life, and the like. Foresters value forests quite naturally in terms of its productivity of merchantable timber. Consequently, the health status of forested ecosystems transformed through harvesting and other means will be assessed in very different ways depending on cultural and social values.

Systems ecologists describe the transformation of complex ecosystems in other terms (such as the four cardinal properties identified by von Bertalanffy). While seemingly value-free, such assessments are far from an exact science. As ecosystems evolve, they display varied structural and functional properties (Rapport and Regier 1992). Under extreme stress, a "one-way" degradation process becomes entrained and an "ecosystem distress syndrome" is commonly found (Rapport, Regier, and Hutchinson 1985). The syndrome may be characterized by the following group of symptoms: reduced primary productivity (except, of course, in moderately eutrophicated aquatic systems); loss of nutrient capital; loss of species diversity; dominance by short-lived, opportunistic, and often exotic species; increased fluctuations in key populations; retrogression in biotic structure (a reversal of the normal successional processes whereby opportunistic species replace species more specialized in habitat and resource use); and increased incidence of disease (Rapport 1989b).

TAKING NATURE'S PULSE

That ecosystems can become severely crippled at the hand of humanity, there is not the least doubt. There have long been observations of local pathologies—some going back at least to Aristotle, who remarked on "small red threats (apparently tubificid worms) growing out of fowl mud" (Leppakoski 1975:2). Today, environmental transformations from human activity have spread from local to regional and now to global levels. (Witness the depletion of the ozone layer, for example, and the increasing concentration of carbon dioxide in the atmosphere leading to the risk of rapid and catastrophic climate change.)

No single disease process has led to the current environmental predicament—unless one speaks of human imprudence. Rather there are a multitude of stressors, many of them interactive, that have caused ecosystems to degenerate. In some cases, a single stressor predominates. (For example, acid precipitation has caused a number of lakes in certain regions of eastern Canada and the United States, as well as northwestern Europe, to lose many valued species (Minns et al. 1990; Harris et al. 1988). But in most regions of moderate to intense human settlement or industrial activity, it is a combination of stressors that deals nature crippling blows. There are many examples of this effect, perhaps the most obvious being the transformation in "big waters," particularly in semi-enclosed seas such as the Baltic and in the Laurentian Great Lakes (Harris et al. 1988).

Yet to read nature's health by exposure to stress is as inadequate as reading human health by exposure to disease. For it is not only exposure but also the innate "resistance" or "susceptibility" of the individual or ecosystem that determines the outcome. Thus the health status of nature (or humans) can be confirmed only by clinical investigation. Depending on one's perspective, such investigations might be directed either to detect symptoms of pathology or to detect the ecosystem's ability to restore itself when subjected to controlled stress.

Taking nature's pulse has proved to be largely illusory. For the vast majority of natural systems, there is scant likelihood of discovering any function remotely equivalent to the pulse in human medicine. Far from exhibiting regularity, most ecosystems might be characterized by arrhythmia—that is, by a dynamics that is highly irregular and punctuated by surprise (Holling 1986). For example, the boreal forest and other perturbation-dependent ecosystems (Vogal 1980) exhibit a very complex dynamics showing marked fluctuations in primary production, species composition, nutrient retention, and insect pest loads.

The problem of detecting disease then becomes one of identifying abnormalities in what is quite naturally a highly irregular pattern and process. This is less a question for theory than for brute empiricism. Here the case method, so well exploited in human medicine, psychology, and business, meets its mettle. (See Rapport 1989b; Harris et al. 1988; Regier et al. 1988.) The surprising fact is that ecosystems which might exhibit much in the way of erratic behavior in the absence of stress may also be resistant to change in the presence of stress (Schindler et al. 1985). A further complicating factor is that many of the signs of ecosystem pathology mimic the natural signs of early successional behavior in perturbation-dependent ecosystems (Rapport and Regier 1992).

In examining the sequence of appearance of signs of pathology, there are divergent patterns. In some ecosystems, the first symptom of distress is registered by losses in "sensitive" species and their replacement by opportunistic, largely exotic, forms. When boreal lakes are artificially acidified, the disappearance of sensitive species precedes any detectable changes in basic ecosystem-level properties such as rates of nutrient cycling or primary productivity (Schindler et al. 1985; Minns et al. 1990). In this case the rates of nutrient cycling and primary productivity remain within normal ranges, even though the pH has been lowered from 6.8 to 5.1 over a seven- year period (Schindler et al. 1985).

In other ecosystems, abnormalities in nutrient export have preceded structural change. Robert O'Neill and colleagues (1977) reported that the first symptom of the effects of toxic substances on terrestrial ecosystems was an increase in nutrient export—before any detectable change in population or community structure. In terrestrial ecosystems damaged by acid precipitation, accelerated leaching of foliar and soil calcium and magnesium were among the earliest symptoms of pathology (Rapport et al. 1985).

This discrepancy suggests that if it is meaningful to speak of the pulse of the ecosystem, different measures are appropriate depending on the nature of the stress and the ecosystem. Fertilization of lakes, for example, may be detected by measuring primary productivity and turnover times of nutrients. For terrestrial ecosystems, changes in community respiration, increases in horizontal nutrient flows, and a reduction in resource use efficiency might be expected to be responsive to certain stresses (Odum 1985). The practical difficulties of accurately measuring such rates of flux for many systems suggest the necessity of looking for structural changes as well—particularly the presence or absence of "indicator" species that suggest the state of the whole ecosystem. Such an approach has a rather long tradition in assessing water quality (Leppakoski 1975, 1979).

The indicator species approach has also been the mainstay in assessing the health of fish communities in the Laurentian Upper Great Lakes. Here "type 1" species are used as a surrogate for ecosystem integrity. Such species have a rather narrow tolerance for conditions deviating from the "somewhat austere conditions originally found in naturally occurring systems" (Ryder and Edwards 1985:39). For oligotrophic aquatic systems, type 1 species comprise lake trout (*Salvelinus namaycush*), deep-water sculpins, and an amphipod (*Pontoporeia hoyi*). Such species have been found to be particularly sensitive to environmental change, and any number of stressors may cause their virtual elimination in local areas. The presence of type 1 species is suggestive of a state of ecosystem health. It is well known, for example, that in both the Baltic Sea and the Great Lakes (Regier et al. 1988) salmonids are sensitive to stresses from nutrient loading, toxic substances, and physical restructuring (dams, diversions, and the like). Yet this same biotic group is relatively insensitive to the loss of near-shore habitat, such as wetlands.

In identifying early symptoms of ecosystem pathology, more attention must be given to critical habitats, which often serve as "centers of organization" for the larger system (Rapport and Regier 1992). In aquatic systems, for example, wetlands, shoals, and the like provide spawning and recruitment grounds for fishes and waterfowl. By harboring "keystone" species (Paine 1980), these regions exert a much wider influence on ecosystem structure than their area alone might suggest (Hilden and Rapport 1991).

MEASURING "ECOSYSTEM FITNESS"

The conventional biomedical model (detection of disease, identification of causes, finding a cure) is in sharp contrast to the salutogenic perspective advanced by Antonovsky (1987). Salutogenic medicine focuses on mechanisms promoting healthiness rather than identifying symptoms or causes of disease. The contrast here is between a diagnostic focus on system *capabilities* (properties of well-functioning feedback systems enabling self-realizing, energetic, and productive processes to occur) and a diagnostic focus on system *disabilities*. In Chapter 14, Robert Costanza distinguishes between a number of definitions of ecosystem health. Defining health as the absence of disease focuses on disabilities; defining health as the potential for ecosystems to recover after perturbation focuses on capabilities. Conventionally, medicine has placed far greater emphasis on correcting disabilities than on fostering capabilities.

What we need in terms of detecting *capabilities* is the ecosystem equivalent of a "knee-jerk" response to stress. I refer here not to the routinely administered doctor's tap on the knee, which, no doubt, has less value as a diagnostic tool than it does as a way to amuse or amaze patients with the magical powers of the practitioner. Rather, I mean the whole area of stress testing in human medicine as a method to detect so-called occult diseases, that is, diseases that are hidden to ordinary inspection. Heart disease in humans, for example, may be masked to common appearances but is readily revealed in tests of pulmonary and lung functions under stress. Similarly, forested ecosystems might to all appearances seem healthy, but their inability to recover after perturbations speaks to the contrary (Bird and Rapport 1986). There are, of course, risks to the patient from administering stress tests. This is particularly the case in patients with impaired circulatory functions. To reduce these risks, less risky tests have been devised. In testing heart function in patients with coronary disease, drugs can be used to mimic the stress effects on the heart of physical exercise, while reducing the risks of heart failure from that very activity.

Stress testing in ecosystems is not entirely a new concept. There have been numerous studies of recovery times for ecosystems under natural stress (see Vogal 1980 for a review of the behavior of perturbation-dependent ecosystems) and induced stress loadings (Bormann and Likens 1979; Bonsdorff 1985). It is here that the notions of "counteractive capacity" (Stebbing 1981), and "resilience" (Holling 1986)—referring to the ecosystem's ability to bounce back after disturbance—come into play.

Clear-cutting the forest is, of course, an extremely invasive way to test the resilience of ecosystems. More benign and carefully controlled stress loads administered on a local scale might serve as well. Goran Milbrink has suggested that the gentle fertilization of Swedish lakes provides valuable information on the capability of the ecosystem to remain healthy. Adding nutrients to healthy lakes tends to increase the number of species in proportion to nutrient concentration, while adding nutrients to lakes already stressed tends to decrease species diversity. As a result, one or a few tolerant species become dominant (Goran Milbrink pers. comm. 1990). For arid ecosystems, an index of "greenness" reflecting the ability to rebound after drought stress serves a similar function. In Australia, this rebound occurs naturally after periods of drought with the arrival of occasional rains in the arid regions (Joe Walker, pers. comm. 1991). The health of these arid ecosystems might be diagnosed by comparing recovery times for known stressed and unstressed ecosystems after periods of drought.

Stress tests in ecosystems do not shed light on all aspects of ecosystem health; nor do the corresponding tests in human medicine. In both cases stress tests ought to be designed to assess fitness by focusing on critical mechanisms that promote a system's integrity. Stress tests in humans concentrate on critical organ functions such as heart and lungs. A corresponding level of specificity must be developed for ecosystems.

CLINICAL ECOLOGY FOR THE 1990s

If the practice of ecosystem medicine is to rest on stronger conceptual and scientific foundations, several basic issues require attention. First, ecologists must devise a more rigorous taxonomy of ecosystem ills, one that can serve as the basis for developing statistics on the success of various efforts at rehabilitating stressed ecosystems. Second, there must be standards to validate treatment options. Establishing criteria by which one treatment might be selected over another requires establishing measures for the success of various treatment options. And third, there must be systematic diagnostic protocols that would enable the ecologist to proceed from detecting general signs or symptoms of ecosystem distress to detecting the major contributory factors to ecosystem pathology.[1]

The specification of a taxonomy of ecosystem ills, the first basic issue, may prove a good deal more complex than the comparable undertaking in the medical field. To begin with, human populations may be far less variable than ecosystems—if for no other reason than they comprise only a single species. Owing to such variability, different ecosystems often exhibit different etiologies with respect to the same stressor (Rapport and Regier 1992). The matter is further complicated by the large number of stressors that impact ecosystems, stressors both of natural and human origin. In view of such complications, present classifications of ecosystem ills (eutrophication, acidification, salinization, erosion, pollution) are fairly primitive. For diagnostic purposes, one needs more specificity—for example, eutrophication in ecosystems of various salinities, mixing times, biotic structure, and so forth.

As in human medicine, the validation of treatments selected for various ills, the second basic issue, is a key requirement for the rational choice of interventions. Relatively little attention has yet been given to this topic with regard to restoration of ecosystem health. The starting point must be a compilation of case studies of efforts to restore ecosystems exhibiting various pathologies. It is here, of course, that the availability of a sufficiently detailed taxonomy of pathologies is crucial to ensuring comparable case history data. Then criteria for success in ecosystem rehabilitation must be established. These might include restoring critical uses made of the ecosystem—for aquatic ecosystems, the criteria for successful rehabilitation might include reestablishing a viable commercial fishery, recreational uses, and drinking water supplies. The criteria should also include various indicators of ecosystem integrity. Once the indicators are selected, it is possible to assess which interventions have worked and which have not—much in the manner that medical treatments are validated. In chemotherapy treatment for breast cancer, for example, success is measured in terms of a period of remission for five or ten years.

When it comes to the development of diagnostic capabilities, we touch upon the third major requirement to advance the practice of ecosystem medicine. The challenge here is to develop diagnostic protocols that begin with generalized symptoms of ecosystem pathology and then proceed to search methodically for additional clues leading to probable causes. In the Lower Great Lakes, for example, the general symptoms of pathology include excessive accumulation of nutrients, loss of valued fish species, and increased population fluctuations (Rapport 1983). For each of these symptoms, there are a number of possible causes. Some could be ruled out by the absence of correlated effects—for example, acidification of aquatic systems not only reduces fish species diversity but also changes the age-class distributions by blocking reproduction and causes shifts in the composition of phytoplankton and zooplankton, eliminating sensitive species. While in certain tributaries to the Great Lakes local symptoms of this nature might be observed, there is no evidence of such symptoms at the level of the whole basins (Lake Erie with its three subbasins and Lake Ontario). Other potential stresses might be ruled in by finding correlated symptoms—for example, eutrophication not only reduces species diversity but also favors increases in the abundance of certain macrozoobenthos, particularly those species with high tolerance for nearly anoxic conditions. In the Lower Great Lakes (Erie and Ontario), there is abundant evidence of this effect for the nearshore and coastal areas (Rapport 1983).

A major effort is required to determine normal ranges for key parameters of ecosystems. Medical practitioners and veterinarians routinely consult published tables giving normal values for key physiological attributes (see, for example, Merck & Co. 1986). Ecologists are hindered precisely by the scarcity of such information. Large-scale ecosystem monitoring efforts began only in the

1970s, encouraged by the International Biosphere Program. This thrust should be continued in order to establish values characteristic of the structure and functions of the earth's major ecosystems (as in Lieth and Whittaker 1975).

One of the major themes of the recently developed Ecological Society of America proposal concerns "sustainable ecological systems." This theme addresses the definition and detection of stress in natural and managed ecosystems, the restoration of such systems, and various management issues. Jane Lubchenco et al. (1991) suggest the need for new research initiatives that address the dynamics and sustainability of large ecosystems under multiple stressors. It is precisely here that holistic concepts of ecosystem health come into play. Advances in clinical ecology will enable more rapid and accurate diagnosis of ecosystem pathologies and point to more efficient ways of restoring damaged ecosystems.

CONCLUSION

Health concepts have become practically ingrained in considering questions of ecosystem breakdown and restoration. Systems science provides a conceptual basis for defining "health." This conceptual foundation must be integrated with social values in order to arrive at scientifically valid but necessarily subjective criteria for ecosystem health. Empirical work has established some of the basic symptoms of stressed ecosystems. As clinical ecology matures, the emphasis should shift away from curative approaches (the focus on ecosystem disabilities) to preventive approaches (with a focus on ecosystem capabilities). In shifting from curative to preventive medicine, a salutogenic perspective suggests the need for noninvasive tests that in effect gauge healthiness. Administering carefully controlled stress tests to ascertain the degree of ecosystem integrity may provide a method for discovering occult diseases in ecosystems well before any overt signs of pathology appear. But many challenges lie ahead. These include designing noninvasive stress tests for critical ecosystem functions as well as developing a better taxonomy of ecosystem ills, more rigorous diagnostic capabilities, and methods to validate proposed treatments or interventions in efforts to restore ecosystem health.

References

Antonovsky, A. 1987. "The Salutogenic Perspective: Toward a New View of Health and Illness." *Advances: Institute for the Advancement of Health.* 4: 47–55.

Bertalanffy, L. von. 1950. "The Theory of Open Systems in Physics and Biology." *Science* 111:23–29.

Bird, P. M., and D. J. Rapport. 1986. *State of the Environment Report for Canada.* Ottawa: Canadian Government Publishing Centre.

Bonsdorff, E. 1985. "Recovery Potential of the Fauna of Brackish-Water Softbotttoms and Marine Intertidal Rockpools." Academic dissertation, Abo Akademi, Finland.

Bormann, G. H., and G. E. Likens. 1979. "Catastrophic Disturbance and the Steady State in Northern Hardwood Forests." *American Scientist* 67(6):660–669.

Brown, L. R. 1990. *State of the World 1990.* New York: Norton.

Dubos, R. 1968. *Man, Medicine and Environment.* New York: Praeger.

Harris, H. J., V. A. Harris, H. A. Regier, and D. J. Rapport. 1988. "Importance of the Nearshore Area for Sustainable Redevelopment in the Great Lakes with Observations on the Baltic Sea." *Ambio* 17:112–120.

Hilden, M., and D. J. Rapport. 1991. "Spatial and Temporal Aspects of Environmental Degradation in the Gulf of Bothnia (Baltic Sea)." Unpublished manuscript.

Holling, C. S. 1986. "The Resilience of Terrestrial Ecosystems; Local Surprise and Global Change." In W. C. Clark and R. E. Munn (eds.), *Sustainable Development of the Biosphere.* Cambridge: Cambridge University Press.

Leppakoski, E. 1975. "Assessment of Degree of Pollution on the Basis of Macrozoobenthos in Marine and Brackish Water." *Acta Acadmise Aboensis* (Abo Akademi), series B, vol. 35, no. 3.

———. 1979. "The Use of Zoobenthos in Evaluating Effects of Pollution in Brackish Water Environments." In *The Use of Ecological Variables in Environmental Monitoring.* Report PM 1151. Stockholm: National Swedish Environmental Protection Board.

Lieth, H., and R. H. Whittaker. 1975. *Primary Productivity of the Biosphere.* New York: Springer-Verlag.

Lubchenco, J., et al. 1991. "The Sustainable Biosphere Initiative: An Ecological Research Agenda." *Ecology* 72(2):371–412.

Merck & Co. 1986. *The Merck Veterinary Manual.* 6th ed. Norwood, N.J.: Merck & Co.

Minns, C. K., J. E. Morre, D. W. Schindler, and M. L. Jones. 1990. "Assessing the Potential Extent of Damage to Inland Lakes in Eastern Canada Due to Acidic Deposition. IV: Predicted Impacts on Species Richness in Seven Groups of Aquatic Biota." *Canadian Journal of Fisheries and Aquatic Science* 47:821–830.

Odum, E. P. 1985. "Trends Expected in Stressed Ecosystems." *Bioscience* 35:419–422.

O'Neill, R. V., et al. 1977. "Monitoring Terrestrial Ecosystems by Analysis of Nutrient Export." *Water Air and Soil Pollution* 8:271–277.

Paine, R. T. 1980. "Food Webs: Linkage, Interaction Strength and Community Infrastructure." *Journal of Animal Ecology* 49:667–685.

Pimm, S. L. 1984. "The Complexity and Stability of Ecosystems." *Nature* 307:321–326.

Porn, I. 1984. "An Equilibrium Model of Health." In L. Nordenfelt and B. Lindahl (eds.), *Health, Disease, and Causal Explanations in Medicine.* Dordrecht: Reidel.

Rapport, D. J. 1983. "The Stress-Response Environmental Statistical System and its Applicability to the Laurentian Lower Great Lakes." *Statistical Journal of the United Nations Economic Commission for Europe* 1:377–405.

———. 1989a. "What Constitutes Ecosystem Health?" *Perspectives in Biology and Medicine* 33(1):120–132.

———. 1989b. "Symptoms of Pathology in the Gulf of Bothnia (Baltic Sea): Ecosystem Response to Stress from Human Activity." *Biological Journal of the Linnean Society* 37:33–49.

Rapport, D. J., and H. A. Regier. 1992. "Disturbance and Stress Effects on Ecological Systems." In B. C. Patten and S. Jorgensen (eds.), *Complex Ecology: The Part-Whole Relation in Ecosystems.* Englewood Cliffs: Prentice-Hall.

Rapport, D. J., H. A. Regier, and T. C. Hutchinson. 1985. "Ecosystem Behavior Under Stress." *American Naturalist* 125:617–640.

Regier, H. A., P. Tuunainen, Z. Russek, and L. E. Persson. 1988. "Rehabilitative Redevelopment of the Fish and Fisheries of the Baltic Sea and the Great Lakes." *Ambio* 17:121–130.

Regier, H. A., and D. J. Rapport. 1991. "Ecosystem Integrity as an Objective in the Great Lakes Basin." Paper presented at the symposium on Defining Ecosystem Health: Science, Economics, or Ethics, American Association for the Advancement of Science, annual meeting, Washington, D.C.

Ryder, R. A., and C. J. Edwards (eds.). 1985. "A Conceptual Approach for the Application of Biological Indicators of Ecosystem Quality in the

Great Lakes Basin." Report to the Great Lakes Science Advisory Board.

Schaeffer, D. J., and E. W. Novak. 1988. "Integrating Epidemiology and Epizootiology Information in Ecotoxicology Studies. III: Ecosystem Health." *Ecotoxicology and Environmental Safety* 16:232–241.

Schaeffer, D. J., E. W. Herricks, and H. W. Kerster. 1988. "Ecosystem Health. 1: Measuring Ecosystem Health." *Environmental Management* 12:445–455.

Schindler, D. W., et al. 1985. "Long-term Ecosystem Stress: The Effects of Years of Experimental Acidification on a Small Lake." *Science* 226:1395–1401.

Stebbing, A.R.D. 1981. "Stress, Health and Homeostasis." *Marine Pollution Bulletin* 12:326–329.

Steedman, R. J., and H. A. Regier. 1990. "Ecological Bases for an Understanding of Ecosystem Integrity in the Great Lakes Basin." *Proceedings of Workshop on Integrity and Surprise*, 14–16 June 1988. Ann Arbor, Michigan: International Joint Commission and Great Lakes Fishery Commission.

Vogal, R. J. 1980. "The Ecological Factors That Produce Perturbation-Dependent Ecosystems." In J. Cairns (ed.), *The Recovery Process in Damaged Ecosystems*. Ann Arbor, Mich.: Ann Arbor Science.

Notes

1. I am grateful to C. Thorpe, Adastra Farm, Napa Valley, California, for suggestions for this chapter.

9

Establishing Ecosystem Threshold Criteria

DAVID J. SCHAEFFER AND DAVID K. COX

Systematic development of ecosystem threshold criteria is a new and significant research area that integrates legal, political, social, and economic disciplines with biological and physical sciences. To address all the complex problems associated with development of threshold criteria, the initial focus must be on a comprehensive synthesis and integration rather than on narrowly defined topics. There are four major research objectives: first, examine the functionality of the proposed definition of an ecological system threshold criterion; second, determine the extent to which such threshold criteria already exist and characterize their properties and units of measurement; third, learn whether meaningful criteria have been developed through the legal, political, social, and economic disciplines and evaluate how well these criteria perform compared with those developed from the physical and biological sciences; and fourth, determine the extent to which ecosystem-condition references can be established.

The legal, political, social, and economic disciplines are being examined to determine whether sound criteria can be developed in the absence of good scientific data. Approaches for developing threshold criteria range from identifying court-imposed criteria to evaluation of consensus criteria to analysis of pending environmental legislation. We are developing new conceptual models that identify the natures and coverages of relationships between the law, economics, and ecosystem sustainability.

In an important review, Stuart Pimm (1984) defined an ecosystem as stable if, and only if, the variables all return to the initial equilibrium after being perturbed. An ecosystem is locally stable if this return is known to apply only certainly for small perturbations and globally stable if the system returns from all possible perturbations. In Pimm's words (p. 322): "The set of all values of the variables from which the system returns to a particular equilibrium is known as the domain of attraction." It is in the context of determining the departure from stability—or, more important, in the context of ensuring the maintenance of stability—that society is now concerned with George Woodwell's question:

> Is it reasonable to assume that thresholds for effects of disturbance exist in natural ecosystems? Or are all disturbances effective, cumulative, and detrimental to the normal functioning of natural ecosystems?
>
> The question is analogous to the classical question of hazards of ionizing radiation and other toxins. If small exposures in addition to background have no discernible effect, but larger exposures do, then it is common to speak of the maximum exposure that has no effect as a "threshold." [Woodwell 1975:10]

The existence of ecosystem thresholds, real or apparent, is assumed (de facto) to provide a practical basis for regulation of pollutants or other stressors such as vehicle traffic:

> If we seek thresholds along [an exposure] gradient, we can find them, but not at the ecosystem level. The thresholds are for survival of species, for the elimination of trees, for the invasion of adventives, for measurable effects on growth. If we measure effects on growth, we find that greater refinement in analysis shows effects at lower concentrations. Any threshold is arbitrary.
>
> If thresholds exist for effects of disturbance, they are few, and

lie at exposure levels below those at which they can be resolved by most current studies of ecosystems per se. [Woodwell 1975:13]

In the following pages we identify several approaches for developing initial threshold criteria. In our models economics appears as a force that modifies legal, political, and social criteria, but is not an independent basis for developing criteria.

DEFINING A THRESHOLD CRITERION

We define a functional ecosystem threshold criterion as: *any condition (internal or external to the system) that, when exceeded, increases the adverse risk to maintenance of the ecological system.* Because of the newness and multidisciplinary nature of this research area, as well as the multiplicity of geographic divisions and types of natural communities, initial research should emphasize general conceptual problems related to criteria development rather than trying to develop site-specific criteria. Regardless of the disciplines and sciences used in this effort, several general issues must be addressed:

• Is our functional definition of an ecological system threshold satisfactory with respect to its ability to be measured?
• To what extent do such threshold criteria already exist? What properties and measurements are used? To what extent, and in what circumstances, are these threshold criteria used? To what extent are they successful in protecting the environment? How is success judged?
• Can meaningful criteria be developed through the legal, political, social, and economic disciplines? How do these criteria perform in comparison with those developed through the physical and biological sciences? Is it more practical (more cost-effective, more rapid, more efficient) to develop criteria through the legal, political, social, and economic disciplines than the biological and physical sciences?
• To what extent can ecosystem-condition reference values be established?

DEVELOPING CRITERIA

Threshold criteria may be developed from five distinct but interrelated categories. In addition to traditional scientific bases, which use results from laboratory and field studies of toxicology and ecology, for example, criteria can be based on legal, political, social, and economic processes.

Legal criteria evolve from three major processes: legislation, administrative rule making, and adjudication. Enabling legislation is enacted by elected (federal or state) legislatures and deals generally with a specific issue. The resulting statute is a compromise between competing interests. The statute scopes the authority for administrative agencies to adopt specific rules. Administrative agencies (the "bureaucracy") promulgate specific regulations based on the general authority of the applicable statute. Again, competing interests can influence the scope or application of regulations. The judicial process responds to challenges of the decision-making process—or to conflicting interpretations of the language of the statute or regulation—by offering its own interpretations of the law or by declaring the law (or part of the law) unconstitutional.

Political criteria are based on the influence of various interest groups in shaping the application of environmental legal requirements. The restrictiveness or permissiveness of legal requirements is based on the resulting compromise between these competing interests. *Social criteria* are determined in various ways. Demographic factors (education, the community's cultural base, wealth, religious influence); population density; the degree, type, rate, and planning of urbanization and development; public sentiment and awareness—all may serve as the basis for social criteria. *Economic criteria* are based on market forces. They are strongly tied to legal, political, and social criteria in that the underlying economic forces will play a substantial role in shaping the compromise that will lead to the establishment of threshold criteria.

The following sections identify two broad approaches to the development of threshold criteria. One approach considers criteria developed through legal, political, and social processes. (We do not consider economics here.) The other considers scientific criteria. To focus the research on achievable endpoints, specific goals should be developed by carrying out a formal "Study Design Assurance" analysis (Schaeffer et al. 1985).

Legal, Political, and Social Criteria

Legal, political, and social influences are significant factors in establishing ecosystem threshold criteria. These influences reflect a prioritization of environmental concerns as they relate to other concerns such as business development and urbanization. At this time, legal, political, and social criteria may be better established, and more easily developed, than technical criteria. Administrative rule making commonly establishes de facto ecosystem threshold criteria, for example, ranging from EPA's promulgation of specific regulations covering an activity or substance to National Park Service policy on the number of people who can use a particular wilderness area each day. Other examples of this type include the issuance of limited numbers of fishing and hunting licenses and imposed limits on the numbers of fish or game taken. Such criteria recognize that the stress which can be tolerated by the ecosystem has limits and, moreover, that the degree of this stress is directly related to population pressures on the ecosystem.

Examination of the development of legal, political, and social criteria should begin with an understanding of how these processes affect criteria. At least four descriptive examinations should be undertaken. First, the process by which federal and state environmental law is developed, and the relationship between them, must be understood. As a matter of constitutional law, federal requirements, to the extent to which they are applicable, are supreme. States do not have the authority to adopt or enforce less stringent requirements. States may only deviate by adopting standards equal to, or more stringent than, the federal requirements. The political influences present on the federal level, or in a particular state, affect the development of environmental law. These competing interests result in compromises that ensure sufficient political support for approval of final requirements. The development of major legislation such as the Clean Air Amendments of 1990 represents compromises by Congress and the president based on industrial and environmental interests. The demographics of the United States or a particular state may provide the basis or "values" underlying environmental requirements. Oregon and Washington, for example, are progressive in adopting requirements that favor the environment. These states are significantly rural, and a large portion of their populations is pro-environment.

Pennsylvania, on the other hand, is a key industrial state with a large percentage of its population directly dependent on industry. As a result, many Pennsylvania state requirements reflect industry positions. By examining the general mission of federal and state agencies involved in resource management and conservation, we can identify governmental agency responsibilities for specific environmental concerns authorized by enabling legislation. Thus a framework for requirements and responsibilities can be organized by subject.

Second, federal and state environmental laws should be examined with multifold objectives. Current statutory requirements that address relevant issues of ecosystem threshold criteria should be identified—for example, use of the Universal Soil Loss Equation to restrict soil loss from farm lands, Department of the Interior restrictions based on usage carrying capacity of primitive areas, and EPA stream requirements for biochemical oxygen demand, dissolved oxygen, nitrogen, and phosphorus, developed from biological site-quality data (Ellis 1937; Richardson 1928).

Third, the technical and policy bases for the statutes and regulations should be examined to identify prevailing interests from which the requirements are derived. The basis for the requirements must also be identified. Then qualitation and quantitation of the criterion and basis must be carried out, as well as identification and analysis of the types of technical information used to establish the criterion. Next, these elements should be ranked or weighted in order to develop a basis for assessing the criticality of each factor to the criterion value. Finally, the relationships of the resulting threshold criteria to existing technical bases must be determined in order to establish their effectiveness, relevance, strengths, and weaknesses.

And fourth, it is necessary to identify the extent to which interactions in the political system result in the development of *meaningful* threshold criteria. The sufficiency of the technical basis must be considered. One must also evaluate the extent to which politics (which is sometimes a negotiated and sometimes a consensual process) can be used to develop technically meaningful criteria when technical data are lacking or there is no technically valid process for developing criteria.

Social factors (specifically demographics) are important determinants of the magnitudes of the stresses received by ecological systems. Consequently, social factors affect both the sensitivity of threshold criteria and the level at which they are established. For example, research is needed to determine the extent to which studies support the hypothesis that the priority for environmental protection increases with education. Another hypothesis to be tested is that the degree of urbanization can be used as an indicator of the health of an ecological system. Under this hypothesis, if system health declines as urbanization increases, it should be possible to use existing ecosystem data to estimate population densities at which adverse changes become identifiable. We might express the population stress on a property system algorithmically as

$$(\text{Property Ecosystem } A/\text{Property Ecosystem } B)^{\pm k} =$$
$$(\text{Population Density } A/\text{Population Density } B)$$

where k represents a stress magnification factor. This expression postulates that system properties such as habitat damage increase—while properties such as numbers of animals and species decrease—with increasing population density.

Scientific Criteria

Development of scientific criteria is a complex undertaking that involves, initially, identification of appropriate biotic measures and, subsequently, controlled testing and refinement. Consequently, both longer time frames and more resources may be needed to develop scientific criteria than legal, political, and social criteria. Initial effort is needed to identify existing scientific criteria, measures that can readily be modified to provide initial criteria, and a methodology and data requirements for developing criteria. Furthermore, each type of ecosystem has its unique species, abiotic components, and interactions, so each type will have its own group of experts. The criterion value for "sufficient nutrient cycling" and methods for its measurement, for example, may differ for each type of ecosystem.

Stressors affecting an ecological system can be broadly classified as physical, chemical, or biological. Off-road vehicles, camping, construction, and a myriad other human activities and (occasionally) natural phenomena exemplify causative factors for physical damage. Acid rain, the ozone hole, oil spills, and industrial effluents are common anthropogenic sources of chemical stressors, while volcanism, sunspots (ultraviolet rays), and global chemical "sinks" illustrate natural chemical stressors. Biological stressors are represented by viral and fungal diseases of plants or animals, introduced species, and perhaps genetically engineered organisms. Although this chapter emphasizes the development of criteria related to the physical and chemical stressors, criteria for biological stressors will be discussed to the extent that they are known.

Scientific criteria are conveniently classified into three categories. *Biological* threshold criteria encompass a wide range of species or biological community responses. The Index of Biotic Integrity, or IBI (Karr 1981; Hite and Bertrand 1989), is computed as the sum of scores for 12 characteristics of the fish community: species richness and composition (total number of fish species; number and identity of darter, sunfish, sucker, and intolerant species; proportion of individuals as green sunfish); trophic composition (proportion of individuals as omnivores, insectivorous cyprinids, and piscivores; number of individuals in sample); and abundance (proportion of individuals as hybrids) and condition (proportion of individuals with disease, tumors, fin damage, skeletal anomalies). Moreover, threshold criteria can be based on the *physical* properties of the abiotic components of the environment, as exemplified by the Universal Soil Loss Equation (USLE) and the Predicted Index of Biotic Integrity (PIBI). The PIBI value is computed from a regression equation relating the IBI to substrate composition percentage of silt/mud, percentage of claypan, mean stream width, and the percentage of open pool area in the sampled fisheries habitat (Hite and Bertrand 1989). Criteria may also be based on the *chemical* properties of the abiotic components. Changes in energy flux relate the abiotic to the biotic components and could be a basis for threshold criteria. The remainder of this chapter suggests some possible approaches for developing scientific criteria.

The State of Illinois and many other states are currently using two fisheries indices—the IBI and the PIBI—and a Macrobenthic Biotic Index (MBI) to evaluate stream quality. Research is needed to determine the applicability of these indices as threshold criteria. There are at least three areas requiring research:

• The sensitivity of the IBI and MBI indices to quantitative changes in the species metrics used in their computation must be evaluated.

• The effects of sampling variation (temporal, spatial) on both the quantitative value of the IBI index and the quality class assignment must be determined either analytically or through simulation studies.

• An appropriate way of using the three indices (IBI, PIBI, MBI) to establish a single threshold criterion must be determined.

Simulation models must be developed to examine the behavior of various threshold criteria. System connectivity models (Levins 1975) appear to be particularly suited for this purpose. For example, alternative analyses of hypothetical systems comprised of a nutrient source, two consumers, and a predator showed that for many of the predictions the detailed structure of the system does not matter (NAS 1981:97–103). For development purposes, various types of threshold criteria can be added to these models and the predictions from the models can be investigated in relation to the criterion. Based on these results, additional components could be added to the model or other hypothetical systems could be devised. Moreover, Lassiter's (1983) ecosystem model should be examined. This model of the effects of chemicals on element cycling and other ecosystem processes includes the use of multispecies representations of biotic communities and mathematical descriptions of the processes that are important in aquatic ecosystems.

Conceptual approaches must be identified and evaluated. Three approaches are possible. First, several properties of ecosystems are critical "to the system's maintenance" (NAS 1981). Those that are easily amenable to measurement, and appear to be particularly relevant, are enumerated by Schaeffer et al. (1988):

• Habitat for desired diversity and reproduction of organisms
• Phenotypic and genotypic diversity among the organisms
• Robust food chain supporting the desired biota
• Adequate nutrient pool for desired organisms
• Adequate nutrient cycling to perpetuate the ecosystem
• Adequate energy flux to maintain the trophic structure
• Feedback mechanisms for damping undesirable oscillations

• Capacity to temper toxic effects, including the capacity to decompose, transfer, chelate, or bind anthropogenic inputs to a degree that they are no longer toxic within the system

If data from an unimpaired ecological system are available as a reference, these properties can be used to devise threshold criteria. In the present context, loss of a species from a prairie habitat (Schaeffer et al. 1990) implies that the habitat has become unsuited for maintaining the "desired diversity and reproduction of organisms." However, loss of a species should be considered an acute, not a threshold, criterion. The properties just listed should be systematically reviewed in the context of current ecological literature to determine if they constitute a suitable basis for devising threshold criteria. Other properties, such as resilience and resistance, have been proposed as measures of ecosystem performance and quality. Pimm (1984, 1991) has presented an excellent review of these properties, but there is no consensus on workable definitions among ecologists. It is this lack of consensus, which is exacerbated by the difficulty of the measurements, that explains the small number of studies cited by Pimm.

Second, use Pimm's (1991) report, the studies he cites, and recent literature to evaluate these definitions and studies as a basis for developing threshold criteria. And third, quantification of multispecies risk could provide a basis for criteria development. The feasibility of developing threshold criteria based on multispecies risk can be examined using models based on the Clarke and Schaeffer (1990) model. One concern is to determine if the existing data are sufficient for developing threshold criteria.

To be meaningful, threshold criteria should be based on quantitative ecosystem exposure-response curves. Although high-quality data from long-term experiments of chemical exposures on ecosystems are limited, a calibrated scale might be developed from studies of radiation effects at the ecosystem level. Radiation effects are directly correlated with chromosome volume, and the nominal and delivered doses of radiation (unlike chemical exposures) are the same. Such a scale could be used to calibrate laboratory hazard assessment bioassays with bioassessment studies using individual, population, and community measures in the exposed ecosystem—a difference in a species' response to radiation in the laboratory and the field, for example, estimates the community's effect on that species.

d: Illinois Environmental Protection Agency, Division of
llution Control.

.M., and D. J. Schaeffer. 1990. "Probabilistic Risk Curves from
zed Exposure and Dose-Response Curves: Application to a
r Radiation Hazard." *The Environmental Professional* 12:114–120.

. G., and G. L. Kingsbury. 1977. *Multimedia Environmental Goals*
vironmental Assessment. Vol. 1. EPA-600/7-77-136a. Washington:

of Engineers. 1990. "Plan of Study: Navigation Effects of the
nd Lock, Melvin Price Locks and Dam." Draft report, February
0. U.S. Army Corps of Engineers, St. Louis District.

M. M. 1937. *Detection and Measurement of Stream Pollution.* Bulletin
. Washington: Bureau of Fisheries, U.S. Department of Commerce.

, R. L., and B. A. Bertrand. 1989. *Biological Stream Characterization*
BSC): A Biological Assessment of Illinois Stream Quality. Special Report
13 of the Illinois State Water Plan Task Force. IEPA/WPC/89-275.
Springfield: Illinois Environmental Protection Agency, Division of
Water Pollution Control.

Karr, J. 1981. "Assessment of Biotic Integrity Using Fish Communities."
Fisheries 6(6):21–27.

Karr, J., K. D. Fausch, P. L. Angermeier, P. R. Yant, and I. J. Schlosser.
1986. *Assessing Biological Integrity in Running Waters: A Method and Its*
Rationale. Special Publication 5. Champaign: Illinois Natural History
Survey.

Lassiter, R. R. 1983. *Project Summary: Prediction of Ecological Effects of*
Toxic Chemicals: Overall Strategy and Theoretical Basis for the Ecosystem
Model. EPA-600/S3-83-084. Washington: EPA.

Levins, R. 1975. "Evolution in Communities Near Equilibrium." In M. L.
Cody and J. M. Diamond (eds.), *Ecology and Evolution of Communities.*
Cambridge: Harvard University Press.

NAS. 1981. *Testing for Effects of Chemicals on Ecosystems.* Washington:
National Academy of Science Press.

Ott, W. R. 1978. *Environmental Indices: Theory and Practice.* Ann Arbor,
Mich: Ann Arbor Science.

Pimm, S. L. 1984. "The Complexity and Stability of Ecosystems." *Nature*
307:321–326.

Pimm, S. L. 1991. *The Balance of Nature?: Ecological Issues in the*
Conservation of Species and Communities. Chicago: The University of
Chicago Press.

Richardson, R. E. 1928. *The Bottom Fauna of the Middle Illinois River, 1913–*
1925. Bulletin 170. Champaign: Illinois Natural History Survey.

Schaeffer, D. J., J. A. Perry, H. W. Kerster, and D. K. Cox. 1985. "The
Environmental Audit. I: Concepts." *Environmental Management* 9:191–
198.

During the 1970s, EPA develo Goals (MEGs) for environmental as contaminants or degradents (in am emissions or effluents conveyed to t. judged to be (1) appropriate for prevent in the surrounding populations or ecolog sentative of the control limits achievab (Cleland and Kingsbury 1977). Although th substances have incorporated few ecological sible to use the concepts included in the MEG proach for establishing ecosystem threshold crit

SCIE
Springfie
Water P
Clarke, A.
Lineari
Nuclea
Cleland,
for En
EPA
Corps
Sec
199
Ellis,
2
Hite

CONCLUSION

The development and use of ecosystem threshold cri important response to newly recognized needs for impre vironmental assessments. In 1992, the State of Illinois initia Critical Trends Assessment Program (CTAP) to determi "state of the State's" ecosystems; results will be used to esta the state's long-term environmental strategies. Initially CTAP examine temporal and spatial patterns of change over sevei decades in "sources" (such as manufacturing and demographics, and "receptors" (defined by the ecosystem critical processes). More detailed examinations will follow using a dichotomous decision key to determine the acceptability of the risk posed by sources to ecosystems. The assessment of an adverse risk can only be made by determining if ecosystem stresses are below thresholds for crit- ical processes. As in human risk assessment, the acceptability of the risk to an ecosystem must be judged in relation to appropori- ate threshold criteria for adverse effects.

References

Bickers, C. A., M. H. Kelly, J. M. Levesque, and R. L. Hite. 1988. *User's Guide to IBI-AIBI—Version 2.01: A BASIC Program for Computing the Index of Biotic Integrity with the IBM-PC.* IEPA/WPC/89-007.

Schaeffer, D. J., E. E. Herricks, and H. W. Kerster. 1988. "Ecosystem Health. I: Measuring Ecosystem Health." *Environmental Management* 12:445–455.

Schaeffer, D. J., et al. 1990. "Ecosystem Health. V: Use of Field Bioassessments to Select Test Systems for Relevant Impact Assessments or Hazard Evaluations: Training Lands in Tallgrass Prairie." *Environmental Management* 14:81–93.

Sokolik, S. L., and D. J. Schaeffer. 1986. "The Environmental Audit. III: Improving the Management of Environmental Information." *Environmental Management* 10:311–317.

Woodwell, G. M. 1975. "Threshold Problems in Ecosystems." In S. A. Levin (ed.), *Ecosystem Analysis and Prediction*. Philadelphia: SIAM.

10

Alternative Models of Ecosystem Restoration

SUSAN POWER BRATTON

Early attempts at restoring natural ecosystems used either an agricultural model or a climax community model emphasizing productivity and stability as indicators of ecosystem health. Current restoration efforts use community structure or ecosystem function models favoring biotic diversity, presence of rare or unique elements, system complexity, and maintenance of natural processes. Naturalness as demonstrated by the self-maintenance of the system or by the reestablishment of elements present before disturbance are also common indicators of health. Present models are inadequate to deal with seminatural landscapes and highly fragmented biotic communities, however, as well as those that have lost keystone species and those that cannot be completely restored. Our concepts of naturalness assume that health is a return to a previous state and do not consider the possibility of new permutations. Naturalness also precludes complex or repeated interactions with human managers. Examples of different philosophical approaches to restoration are provided, primarily, from projects in U.S. national parks.

ROOTS OF RESTORATION

Although humans have practiced ecosystem restoration ever since farmers recognized that allowing former fields to succeed to forest would restore fertility to the soil, the roots of the present practices of ecological restoration lie deep within the conservation and preservation movements of the nineteenth and early twentieth centuries. The two major thrusts of early restoration efforts were regaining the "natural balance" lost by ecosystems disturbed by greedy, abusive Euro-Americans and returning badly degraded lands to a more productive state. The ability of lands and waters to produce either crops or other amenities, such as shellfish and deer, is probably one of the most venerable ways of assessing ecosystem health. Just as one wing of twentieth-century environmentalism has been dedicated to conservation—the wise use of natural resources—one of the major motives for environmental restoration has been to restore degraded lands to productivity. The bountiful harvest of softwoods, clover, or ducks then indicates the ecosystem is healthy. Economic value or lack of it is thus often an important indicator of health.

During the early decades of the twentieth century, ecologists began to emphasize and quantify processes of ecological change, particularly the process of succession. Plant ecologist Frederick Clements viewed natural succession as leading to a climax community, which, when established, was the stable and final biotic assemblage occupying a particular site. According to McIntosh (1985:81), "The climax *association* was, barring external disturbance, a stable and self-reproducing collection of populations, which, in Clements' scheme, was the culmination of the developmental sequence, or seres, from which it came." Climax theory implies that the final community in the successional sequence has the most biomass, the most complex nutrient cycles, the greatest productivity, and the greatest species diversity. If freed from major disturbance, it also is able to maintain itself indefinitely. To a great extent, early efforts at restoring forest, prairie, and wetland vegetation reflected the notion of the climax (and, perhaps covertly, nineteenth and early twentieth century social ideals).

A typical early restoration effort is the Civilian Conservation Corps' planting of white pine (*Pinus strobus*) on former agricultural fields in the southern Appalachians. There was a recognition that the sites were no longer useful for agriculture but might be returned more quickly to a forested state if the early stages of succession were skipped. White pine is a valuable and relatively fast-growing timber species. It is found in old-growth forests in the region but can also occupy open sites. Starting a mixed hardwood forest would be far more difficult and no more economically productive. The sites might be maintained for years in white pine, via culling, harvest, and replanting, or allowed to revert to native cove and oak forest. When the plantings were done during the New Deal, there was little concern for stand diversity, long-term successional trends, or the state of the forest understory. The goals were soil recovery, timber production, and site stability. A dense stand of white pine will, in fact, inhibit reestablishment of many forest floor herb species that would naturally accompany deciduous forest succession on the same sites.

Some of the pine plantings were in what are now Great Smoky Mountains National Park and the Blue Ridge Parkway. There was no concept at the time of restoring the "natural" forest structure, other than by means of unmanipulated native plant succession. It is also notable that in the early days of the Great Smoky Mountains National Park the only mammal evaluated for restoration was the white-tailed deer, which had been reduced by overhunting to a small herd in an inaccessible valley called The Hurricane.[1] The timber wolf, eastern mountain lion, river otter, and fisher were not considered possible candidates for restoration until the 1970s and 1980s.[2]

A second, and even more telling, example of early restoration efforts was the project on Cape Hatteras National Seashore to establish an artificial dune line. The Outer Banks of North Carolina overwash frequently during hurricanes and northeasters. The natural dune and shoreline are not geographically stable and migrate relative to geophysical variables, such as changes in sea level and storm frequencies. Since successional theory suggests that stable dune lines lead to development of more stable plant communities behind them, land management agencies viewed establishment of a high and very stable fore dune as a way to prevent the sea from flooding roads and buildings. The State of North Carolina, the Civilian Conservation Corps, and the National Park Service collaborated in the construction of a high fore dune (sand on brush piles or fences) topped with a dense sward of nonnative, American beach grass (*Ammophila brevigalata*). The unidirectional change of climax theory is ill applied to a barrier island system that is slowly moving landward, however. Although the beach grass did temporarily stabilize the dunes, the Atlantic is slowly encroaching on their bases. The shrub communities that have developed behind the dunes represent a type of vegetation usually found on higher ground, well outside the reaches of salt spray. They will almost certainly never reach the "stable" forested state because, sooner or later, an overwash event will kill them with salt and standing water (Dolan et al. 1973; Schroeder et al. 1976). Ironically, the shrub communities tend to block the view of both the dunes and the marshes, leaving the Outer Banks tourist driving through a tunnel of woody vegetation dependent on continuing human interference with coastal hydrology. The goal of advancing succession has failed to accommodate the dynamics of island ecosystems that support numerous disturbance-tolerant species.

NATURAL HARMONY

In contrast to the manipulative and productivity-oriented mode of ecosystem restoration, there were those who favored protection for native ecosystems and generally thought the best way to restore native biotic communities is to keep people and their activities out. This camp was also influenced by notions of natural ecosystem "balance" and the ideal of a natural climax community. John Muir, for example, saw loss of wildness or loss of system integrity as undesirable states. In *My First Summer in the Sierra*, Muir (1911) recognizes that sheep cause major changes in the flora of high-elevation meadows and laments that "almost every leaf that these hoofed locusts can reach within a radius of a mile or two from the camp has been devoured." He notes that although the sheep cannot kill the trees, some of the seedlings are damaged—and if the herds become large enough, "the forests, too, may in time be destroyed." Muir cynically predicts, "Only the sky will then be safe, though hid from view by dust and smoke, incense of a bad sacrifice." The same themes appear in his petitions for the establishment of Yosemite National Park (Muir 1970:56), where he complains of the lumber men and shepherds and suggests that upper-elevation watersheds as well as the areas surrounding the big tree groves be included in the park:

For after the young, manageable trees have been cut, blasted and sawed, the woods are fired to clear the ground of limbs and refuse, and of course seedlings and saplings, and many of the unmanageable giants [trees] are destroyed, leaving little more than black charred monuments. These mill ravages, however, are small as yet compared with the comprehensive destruction caused by the "sheepmen." Incredible numbers of sheep are driven to mountain pastures every summer, and desolation follows them. Every garden within reach is trampled, the shrubs are stripped of leaves as if devoured by locusts, and the woods are burned to improve pasturage. The entire belt of forests is thus swept by fire, from one end of the range to the other; and with the exception of the resinous *Pinus contorta*, the sequoia suffers the most of all. Steps are now being taken towards the creation of a national park about the Yosemite, and great is the need, not only for the sake of the adjacent forests, but for the valley itself.

In his petition concerning the preservation of the Hetch Hetchy Valley along with the Yosemite Valley, Muir (1970:97) proposes "ax and plow, hogs and horses" as agents of disturbance in "Yosemite's gardens and groves" and states in conclusion: "And by far the greater part of this destruction of the fineness of wildness is of a kind that can claim no right relationship with that which necessarily follows use." Muir's basic strategy for restoration was to get the people out or control their activities.

This concept of the value of naturalness came one step further and began to intrude on the notion of productivity in Aldo Leopold's work. Leopold (1949:130–131) is speaking against productivity as the best indicator of ecosystem health when he writes of wolf eradication accompanied by protection of deer:

I have lived to see state after state extirpate its wolves. I have watched the face of many a wolfless mountain, and seen the south-facing slopes wrinkle with a maze of new deer trails. I have seen every edible bush and seedling browsed, first to anemic desuetude, and then to death. I have seen every edible tree defoliated to the height of a saddlehorn. Such a mountain looks as if someone had given God a new pruning shears and forbidden Him all other exercise. In the end the starved bones of the hoped-for deer herd, dead of its own too-much, bleach with the bones of the dead sage, or molder under the high-lined junipers.

This description is not unlike Muir's description of sheep as armies of antienvironmental terrorists. Leopold, however, does not see the problem as solely humans and their livestock attacking nature. He also sees it as wild nature damaging wild nature when a critical element of the "land community" is extirpated. The goal of increasing the productivity of the wild, when taken to extreme, is little different than direct degradation by humans. Leopold (1949:132), in fact, equates too many cows with too many deer:

> I now suspect that just as a deer herd lives in mortal fear of its wolves, so does the mountain live in mortal fear of its deer. And perhaps with better cause, for while a buck pulled down by wolves can be replaced in two or three years, a range pulled down by too many deer may fail replacement in as many decades.
>
> So also with cows. The cowman who cleans his range of wolves does not realize that he is taking over the wolf's job of trimming the herd to fit the range. He has not learned to think like a mountain. Hence we have dustbowls, and rivers washing the future into the sea.

Leopold changed the temporal perspective and recognized the difference between short-term disturbances, such as predation on the deer herd, and long-term changes in community structure. Further, Leopold recognized that the absence of a single keystone species may entirely modify the way an ecosystem functions.

When Leopold developed his land ethic, he used the concept of the ecological community, which incorporates "soils, waters, plants, and animals," and declared that "conservation is a state of harmony between men and the land" (p.204). Leopold essentially established an ethic for management where humans need to consider both the value and function of all members or parts of an ecosystem. This model has been very influential in contemporary restoration, where productivity may play second fiddle to other values such as "naturalness" and recovery of biotic diversity. Leopold emphasized the presence of all the original parts of a system, not just the harvestable white pines and deer. Further, he valued the soil and water for their own sake, not merely as abiotic factors necessary to plant growth.

Leopold was directly involved in the establishment of the University of Wisconsin–Madison Arboretum, which began experimental restoration of native plant communities during the 1930s. "The time has come," Leopold argued, "for science to busy itself with the earth itself. The first step is to reconstruct a sample of what we had to begin with" (Jordan et al. 1987:7). As with his efforts in wilderness preservation, Leopold was looking back at the frontier and trying to protect or replicate what had not yet been plowed under. The early restoration projects at the arboretum emphasized the vanishing prairies, and coincidentally, after the first efforts failed to produce a viable native grassland, aboretum scientists began experimental management utilizing fire as a necessary disturbance. The successful establishment of the Curtis Prairie proved that fire was part of the "natural balance" or "natural cycle" and further convinced ecologists that maintaining all the parts of the ecosystem was necessary to maintaining the whole.

CONTEMPORARY PERSPECTIVES

The contemporary restorationist, particularly someone who works with wetlands, prairies, or other native ecosystems, is likely to hold "naturalness" as a major goal, in terms of both ecosystem composition and function. A restoration might be considered successful if all the species originally present on the site have returned, if it has high diversity or the original diversity of species, or if natural disturbance cycles (such as natural fire regimes) have been reestablished. Further, there is a current concern for the inclusion of rare or unique elements, so the understory orchids, forgotten in the white pine plantings of the 1930s, might today receive a fair amount of managerial time during a restoration effort. Worry over the potential extinction of thousands of species worldwide has placed an extra emphasis on endangered species, and in some cases restoration efforts may actually be dominated by "recovery plans" for endangered flora and fauna. In federal land management there has been a shift from managing for the most productive, consumable, or, in the case of the National Park Service, "dramatic" elements of the system toward managing for the most limited, unique, or fragile elements.

Behind this philosophical orientation is a strong concept of "natural balance" and the necessity of keeping all the parts of a system, even if they look insignificant to us. Walter Westman (1991:210), in a paper discussing potential ecological criteria for evaluating the success of a restoration project, has observed: "A common implicit goal of ecological restoration is to restore a self-sustaining, homeostatic system whose component species have a long coevolutionary history." One could thus define an ecosystem once "sick" but now restored to "health" as: self-replicating or self-sustaining; characterized by or developing toward homeostasis (perhaps undergoing repeated disturbance but returning to a similar state); having biotic components that are interrelated and occur in established associations.

A restoration project typical of contemporary philosophy is the major effort made to restore, not just the forest cover, but the actual topography of the Redwoods National Park during the late 1970s and early 1980s. Logging in the giant redwood stands had more than removed trees; it had also modified hydrology and initiated slope erosion. As well as attempting to establish new vegetation cover as quickly as possible, the National Park Service utilized heavy earth-moving equipment to obliterate roadbeds and reestablish the predisturbance contour of the slopes. The park incorporated sites on serpentine bedrock that are very liable to soil erosion and landslides. The restoration project gave high priority to these serpentine grasslands (which are also known for their endemic flora), reconstructed the slopes and reestablished native species. The project recognized, first, that logging-initiated disruption of hydrological cycles might encourage floods and higher than normal water levels on the streams, which in turn might harm the remaining unlogged redwoods (Agee 1980). Further, the logging roads were potential sources of ongoing erosion and drainage problems, as well as being potential long-term scars in the landscape. Although the National Park Service did not (and could not) attempt to bring the redwood forest immediately back to its former glory, they made a major effort to return the substrate to its prelogging condition and thereby to encourage homeostasis. The Great Smoky Mountains National Park, in contrast, made no attempt, after the heavy logging of certain watersheds in the 1920s and 1930s, to remove the skidder trails or accelerate the return of the cove forests.

Another example of contemporary restoration philosophy is the trend toward reestablishing predators in national parks. Although some of the species concerned are endangered, many are not. Cumberland Island National Seashore, for example, is in the process of reestablishing a bobcat population. The bobcat is neither endangered nor rare in the region. The species occurred on the island historically and is one of several predators extirpated by Euro-American occupation of the site. The public appears to accept the concept of restoring bobcats "just because they belong there." Managers, however, are concerned about intensive browsing of vegetation by native deer and the proliferation of nonnative wild hogs and armadillos. Thus the bobcat is a means of not just putting a predator, but also "predation," back into the island ecosystem (Warren et al. 1990). The return of the eastern mountain lion, the wolf, and the black bear remain more controversial and technically problematic.

John Cairns (1991:186) says that "restoration means recreating both the structural and functional attributes of a damaged ecosystem." Westman (1991:204), after noting that structure and function cannot always be restored simultaneously, suggests monitoring structural characteristics and functional characteristics to determine restoration "performance." He proposes evaluating structure by comparing species composition (or some measure of species importance) with the original species composition or with "regional baseline levels"; he suggests measuring function by using probabilistic (Markov) models to determine the likelihood of transition from one state to the next and monitoring the resilience of the ecosystem. The latter is rather like stress testing in humans. If the system can recover easily from perturbations, shows little deviation from natural successional processes, and is well damped (does not vary greatly from the expected degree of oscillation), the system is healthy.

LIMITS ON NATURAL RESTORATION

In reviewing contemporary trends in restoration, two important values repeatedly appear. First, current restoration practice tends to look to the past. It attempts to replicate previous conditions or to preserve disappearing ecosystems. Second, it places a high value on naturalness and the establishment of biotic communities that can survive without further human care. Often naturalness is interpreted as "natural balance." In many circumstances, these values are perfectly appropriate and one can only applaud the University of Wisconsin for the work it has done in preserving and restoring native American grasslands. Yet these values are also limiting. They create a dichotomy not only between "natural" and "human" but also between "natural" and "new." Our mental models and ideals may actually be constraining the ways in which we use ecological restoration techniques to solve environmental problems. Our concept of ecosystem health emphasizes "natural" homeostasis and past ecosystem states.

Holding the "natural past" as the ideal ignores a number of limitations that commonly affect restoration projects. Although Westman (1991) notes that "defining the historical condition of a habitat frequently is possible only in general terms," many attempts at determining restoration performance are based on poorly known previous system states. The use of predictive models, even probabilistic models, must either be based on an accurate knowledge of the structure and function of an ecosystem prior to the restoration effort (and prior to the disturbance that is prompting the restoration), or it must be based on a general analysis of the structure and function of the desired ecosystem type. In the field, one is not always lucky enough to know what an area was like before major anthropogenic disturbance. Further, species associations tend to be rather plastic. The fact that a specific orchid was present before the critical perturbation does not mean that the species would have remained at the site for the next fifty years if that perturbation had not occurred.

Not only is it difficult to know what an ecosystem was like, but it is also hard to know where it is headed, particularly in light of global change. In the example from Cape Hatteras cited earlier, the islands had already been heavily grazed by livestock before the artificial dune line was constructed. Therefore, the pre-dune line and pre-park state of the systems was not "natural" and we do not know exactly what vegetation would have been present if the islands had not been grazed. If cattle and ponies extirpated a rare plant species, we have no record of it. If we wished to remove the dune line and restore the vegetation, we cannot assume the shorelines will be static. We do not know with any certainty whether global climate change will cause a change in storm frequency or an increase or decrease in precipitation.

Further, many areas slated for restoration are no longer integrated with surrounding ecosystems in terms of their original landscape context. The area to be restored may only be part of a once far more extensive marsh, or it may be one of several high alpine areas in a mountain chain. The restored area may now be isolated from similar systems. Elements may be missing, and they may be difficult to replace. Large predators, with the exception of the alligator and the recently reintroduced bobcat, are gone from Cumberland Island National Seashore, for example. One could bring in a couple of wolves or a panther or two, but this will not establish a viable population if there are no breeding, wild populations of wolves or panthers on the mainland or on the other adjoining islands.

Past human activities may have modified the geophysical environment to such an extent that certain previous residents of the native biota can no longer survive on the site. Disturbance regimes may also have been altered. The construction of a dam, for instance, changes flood frequency and intensity, as well as patterns of silt deposition on the floodplain. The presence of roads in an area may increase the occurrence of anthropogenic fires along the rights of way, while preventing small lightning-ignited blazes from crossing the highways. Even without direct human support and encouragement, nonnative species may have invaded an area and may be difficult to suppress or completely eliminate.

We often assume that "natural climax" ecosystems will be the most resilient, but this may also be a matter of scale. A patch of kudzu (*Pueraria lobata*), the fast-growing vine that often takes over roadsides and abandoned fields in the southeastern United States, is very resilient when disturbed (it is extremely difficult to eradicate), and the species inhibits native plant succession and prevents trees from reoccupying the site. Kudzu rarely penetrates large continuous areas of mature deciduous forest, yet in a fragmented landscape kudzu jungles often become one of the most stable features. Resilience is desirable in a restored ecosystem, but "most resilient" may not in turn mean "natural balance."

Contemporary restoration efforts are beset by a variety of scale and boundary problems (even though landscape and ecosystem boundaries may be difficult to define). The concept of "naturalness" may itself be susceptible to changes of scale. As Bryan Norton et al. (1991) suggest, all systems are, in the end, dynamic, and maintaining a reasonable degree of system integrity at the large-scale level requires maintaining a reasonable degree of system integrity at the landscape or small-scale level.

Restorationists themselves are fully aware of these difficulties and recognize that the restored ecosystem may be anything but static. Grant Cottam (1987:263), in fact, describes the long-term changes documented by a series of vegetation surveys in the restored prairies at the University of Wisconsin Arboretum and notes:

> Two characteristics are important: the first is that a certain amount of shifting in the distribution of community types does occur during the 25 years between the first and last survey, as introduced species find their way to optimum sites; the second, however, is that there is also a great deal of shifting that cannot be accounted for in this way. In fact, the various communities do not have static borders, but appear to change location with every survey in an amoeba-like movement that seems to be a response to the short-term climatic events of the years immediately preceding the survey.

Restorationists have also been accepting compromise with completely "natural" restorations. One of the best examples of this is the move toward native "lawns" and plantings around developments as part of the landscape architect's repertoire. Native prairie species have been established around office buildings and suburban homes. The more adventurous efforts even incorporate burning of the native grasses, rather than mowing, to suppress shrubs and unwanted nonnative species.

FUTURE RELATIONSHIPS

Despite the progress made by more imaginative restorationists, our "natural past" model of restoration and ecosystem health still limits how and where restoration techniques are applied. We desire either a "natural ecosystem" or one that is productive or intensively managed and have little concept of anything in between. This means we either demand a national park or a cornfield. It is notable that in Europe over the centuries, seminatural woodlands and meadows evolved that were both productive and rich in native flora and fauna. English woodlands dating back to the Middle Ages and perhaps even to Celtic times often retain a diverse flora of understory forbs (Peterken 1981). Hay meadows cut and fertilized on traditional rotations are often species-rich, as are many managed heath communities. Ecological reality confirms that seminatural communities and those maintained by continuing human interaction can help to preserve biotic diversity and other "natural values," particularly where the landscape is already intensively managed for human ends. (See Callicott 1991 for a critique of the separation of the natural and the human.)

We must take a careful look at how we evaluate human manipulation of systems relative to their health. We tend to view no manipulation as good and manipulation optimizing productivity as good and states in the middle as undesirable. Most of us recognize that cultures with long histories of interaction with local ecosystems often live "in balance" with those ecosystems. Rather than interpreting this as a healthy human/nature relationship, however, we often portray these cultures as representing "natural men"—that is, humans who have remained part of nature. Such situations are thus easily absorbed by the "natural past" model of health. This may also lead to a misunderstanding of how non-Western cultures have interacted with the landscape, and we may assume these cultures have been less liable to change than has actually been the case. An obvious example is the impact of the Native American agriculturist on New World ecosystems. Just as in Europe, new farming techniques were developed, agricultural areas expanded, and populations increased. There was never a magic point where humans found a perfect "balance" with the surrounding vegetation and wildlife. Climate change, depletion of game populations, and other factors may, in fact, have stressed or displaced Native American cultures throughout their occupation of the two continents. This suggests several conclusions:

• The degree of human manipulation of a "healthy" ecosystem depends on the landscape context. We might continue to consider constant human intervention undesirable in a national park, while accepting a moderate level of manipulation in woodlots surrounded by farm fields. We have to evaluate the impact of human manipulation at all ecosystem scales to fully discern "beneficial" and "unbeneficial" human interactions.

• Homeostasis is an unrealistic goal and restoration efforts should assume some change in species composition and abiotic factors through time. Again, restoration should be placed in its landscape context. The biological integrity of a single site cannot be determined by a rigid list of necessary components, yet a large number of extinctions within a region or biogeographic province indicates that landscape change is occurring more quickly than the native biota can accommodate the disturbances.

• In some cases, new combinations of assemblages of species may be acceptable, or species may occupy new habitats. Although there are certain locations and ecosystem types where this may not be desirable, there are others where this strategy may be the only option.

• We need to understand how systems change—and then establish pragmatic goals concerning acceptable levels of change.

Among the dilemmas of natural area management, there are numerous situations where a broader view of either acceptable system change or the acceptable degree of human interaction might be desirable. In the southern Appalachians, for example, endangered rock outcrop plants are being trampled by rock climbers and scramblers. Some species, such as the southern Appalachian mountain avens (*Geum radiatum*), which grow on wide ledges, have probably lost entire populations to human ignorance. As the mountain avens is found on widely scattered cliff faces and rocky peaks, each population may have some genetic value in maintaining the species as a whole. As part of a restoration effort at Craggy Pinnacles on the Blue Ridge Parkway, North Carolina, greenhouse-raised mountain avens plants, from local genetic stock, were reestablished on disturbed rock ledges. (This and the following examples are from personal field experience and managerial contacts.)

Since the mountain avens population at Craggy Pinnacles is very limited in size, the suggestion was made that some of the artificially germinated plants be placed on man-made road cuts below the natural cliff faces. This is not the original habitat for the avens, nor is it a "natural" habitat. The road-cut plants, however, could serve as "genetic backup" should something happen to the population on the Pinnacles. Further, the plants on the road cuts would have a high-elevation environment, while any plants "stored" at a botanical garden would grow up in milder temperatures.

After careful evaluation of the microhabitats on the natural cliffs and on the road cuts, the National Park Service successfully established mountain avens on ledges created by blasting. Ironically, this portion of the Craggies project remains controversial because the Park Service does not wish to write an environmental impact statement for an artificially established rare species population each time they need to do maintenance work on the road shoulders and tunnels.[3] The U.S. Fish and Wildlife Service has wavered on whether an impact statement will be necessary. Yet road cuts could provide additional habitat for several other high-elevation outcrop species that are presently threatened or endangered, while colorful flowering plants, like the mountain avens, would greatly improve the appearance of bare rock faces. Moreover, road cuts are reasonably free of technical climbers and scramblers.

Someone interested in rare species protection might object at this point that establishing new populations, even if one uses local genetic stock or stock that cannot interbreed with other "natural" populations, invites destruction of existing endangered species habitat. If a careful legal distinction is made between the naturally established populations and those that are anthropogenic, and the latter are carefully established and managed, these difficulties will be minimized. We also must ask ourselves about the consequences of managing the mountain avens site by site, rather than as a series of populations in the southern Appalachian landscape. There may be two or more relevant scales for approaching the preservation of not just the mountain avens as a species but the genetic diversity contained within the various mountain avens populations. (See Norton et al. 1991 for a theoretical discussion of resolving scalar/hierarchical problems and Noss 1991 for a discussion of restoration of entire wilderness areas.)

A second example is that of manatees using Georgia paper plant effluents which provide warm water and act as artificial warm springs. The manatees' use of the sites is two-edged because the paper plants pose certain threats to the manatees as well as providing a haven from the cold. The plant effluents, of course, may contain traces of toxins. And if the manatees stay around the effluents during the winter, they run the risk of being frozen to death if the effluent stops running or there is a sudden drop in temperature. Although they risk hypothermia in Georgia, manatees sometimes experience food shortages in Florida and the Georgia marshes provide a steady, if hard to reach, supply of saltmarsh hay (*Spartina alterniflora*). If we wish to protect manatees, then we should either try to stop them from using the effluents, and order them back to Florida, or else make the effluents as safe as possible for the manatees. "Safe" or "healthy," in this case, means designing an effluent as ecologically similar as possible to a natural spring, with the additional concern that the site be protected from boat traffic. Recent proposals for management of the effluent at one Georgia mill include developing a deep basin for the manatees, continuing warm-water flow at all times (particularly at low tide, when manatees are less likely to be feeding away from the effluent), and preventing entry by boats. This, then, is a new habitat on an "old natural" design.

A third example is that of the bobcat restoration and similar efforts to reestablish species on islands or in reserves of limited areas. If we decide we always must have a population that meets genetic viability criteria for minimum population size, then many small areas cannot support larger vertebrate predators. Larry Harris and others (Harris 1988; Harris and Gallagher 1989) have suggested that establishment of corridors will relieve this problem, but this is not always possible. Our "natural past" model inhibits us from accepting the fact that humans may have to intervene occasionally and manage the gene pool—perhaps by arranging a few "accidental" introductions of new blood. Bobcats can swim from the mainland to Cumberland Island and vice versa, but present coastal development patterns undoubtedly slow the rate of this exchange. Some genetic management may ultimately be necessary to prevent inbreeding depression on an island that can probably support only twenty-five to forty bobcats in the long run. Maintaining genetic "health" in this case may require human intervention.

A fourth example is the many Piedmont and Appalachian forests that have probably lost part of their understory herb species due to repeated cutting and other disturbances. We do not usually think about restoring the wildflowers on these sites, especially if they continue to be utilized for timber or they surround human suburbs. Yet, especially in the latter case, the restoration of the missing flowering species would add greatly to the beauty and biotic diversity of the stands. Further, the stands may be isolated

from each other, and native forest floor species may have difficulty dispersing among widely separated sites. When the forests
have regrown and the soils have recovered their litter layer, the
climatic conditions within the stand may again be favorable for
the still absent forbs. In this case, we must recognize that the return of the canopy does not mean the return of all native biota,
and that with fragmentation of the surrounding landscape matrix,
some of the more slow-growing species may be slow to reinvade
without human assistance.

Another case concerning habitat loss and fragmentation is the
destruction of the small meanders and the oxbow wetlands on
Abrams Creek in Cades Cove, in Great Smoky Mountains
National Park. The Soil Conservation Service suggested channeliz-
ing the stream, which runs through a historic district, to reduce
erosion. The channelization resulted in little erosion control and
seriously disturbed rare plant habitat along the banks and in the
floodplain. A small wet meadow remained near the end of the
cove, and several species of state concern were collected from this
locality. Traditionally this meadow had been mowed. In recent
years, however, mowing has been abandoned (the area is too wet
for modern tractors) and shrubs have begun to displace forbs and
grasses. Although mowing is a human activity, one could not say
it was "unhealthy" for the site for it helped to maintain a high diversity of early successional and floodplain species, some of which
had lost habitat during the channelization.

A last example takes us back to the artificial dune line at Cape
Hatteras. Recently, the National Park Service has been experimenting with fire to temporarily control the shrubs encroaching
on the road margins and wetland edges. The resulting plant
communities are not those that would be found after overwash,
nor are they those encouraged by the high, beach-grass-covered
dune. The philosophical question then becomes: When one cannot
suddenly remove the dune, because roads and developments will
be more frequently flooded, is the pyric community acceptable as
a substitute for the overwash community? One might argue, on
the one hand, that the system will never be truly healthy until the
natural overwash processes have been restored. Disaster is always
looming just around the corner when natural coastal processes are
disrupted. But the human coastal communities have not yet coped
with the dilemma, and fire represents an easily managed distur-

bance that suppresses shrubs and results in native plant associations.

CONCLUSION

Our "natural past" model for determining the success of a restoration effort limits the ways in which we apply restoration techniques. It would be better, considering the number of threats to native species, to stratify and maintain "naturalness" and "homeostasis" as goals for certain sites while accepting some degree of "seminaturalness" and "human relationship" for others. We should also decouple "productivity" as a criterion from "unnatural" and assume that "seminatural" sites can maintain high productivity and high biotic diversity if humans manage them properly. Our present views of naturalness often ignore problems of scale, as well as possibilities for ecosystem management or restoration that will help to compensate for changes at both large and small scales. Ecosystem restoration is a relatively new field, and its applications are continually expanding. Our attachment to the "natural past" model of restoration may be causing us to miss good opportunities and to respond insufficiently to the ecosystem change and habitat fragmentation that is presently occurring worldwide.

References

Agee, J. 1980. "Issues and Impacts of Redwood National Park Expansion." *Environmental Management* 4(5):407–423.
Cairns, J. 1991. "The Status of Theoretical and Applied Restoration Ecology." *The Environmental Professional* 13(3):186–194.
Callicott, J. B. 1991. "The Wilderness Idea Revisited: The Sustainable Development Alternative." *The Environmental Professional* 13(3):235–247.
Cottam, G. 1987. "Community Dynamics on an Artificial Prairie." In W. R. Jordan, M. E. Gilpin, and J. D. Aber (eds.), *Restoration Ecology: A Synthetic Approach to Ecological Research.* Cambridge: Cambridge University Press.

Dolan, R., P. J. Godfrey, and W. E. Odum. 1973. "Man's Impact on the Outer Banks of North Carolina." *American Scientist* 61:152–162.

Harris, L 1988. "Landscape Linkages: The Dispersal Corridor Approach to Wildlife Conservation." *Transactions of the North American Wildlife and Natural Resources Conference* 53:595–607.

Harris, L., and P. B. Gallagher. 1989. "New Initiatives for Wildlife Conservation: The Need for Movement Corridors. In *Preserving Communities and Corridors*. Washington: Defenders of Wildlife.

Jordan, W. R., M. E. Gilpin, and J. D. Aber. 1987. "Restoration Ecology: Ecological Restoration as a Technique for Research." In W. R. Jordan, M. E. Gilpin, and J. D. Aber (eds.), *Restoration Ecology: A Synthetic Approach to Ecological Research*. Cambridge: Cambridge University Press.

Leopold, A. 1949. *A Sand County Almanac and Sketches Here and There*. London: Oxford University Press.

Linzey, A. V., and D. W. Linzey. 1971. *Mammals of Great Smoky Mountains National Park*. Knoxville: University of Tennessee Press.

McIntosh, R. P. 1985. *The Background of Ecology: Concept and Theory*. Cambridge: Cambridge University Press.

Muir, J. 1911. *My First Summer in the Sierra*. Boston: Houghton Mifflin.

———. 1970. *The Proposed Yosemite National Park—Treasures & Features*. Olympic Valley, Calif.: Outbooks. Originally published in 1890.

Norton, B. G., R. E. Ulanowicz, and B. D. Haskell. 1991. *Scale and Environmental Policy Goals*. A Report to the EPA., Office of Policy Planning and Evaluation. Washington: EPA.

Noss, R. F. 1991. "Wilderness Recovery: Thinking Big in Restoration Ecology." *The Environmental Professional* 13(3):225–234.

Peterken, G. F. 1981. *Woodland Conservation and Management*. London: Chapman & Hall.

Schroeder, P. M., R. Dolan, and B. P. Hayden. 1976. "Vegetation Changes Associated with Barrier-Dune Construction on the Outer Banks of North Carolina." *Environmental Management* 1(2):105–114.

Warren, R. J. et al. 1990. "Reintroduction of Bobcats on Cumberland Island, Georgia: A Biopolitical Lesson." *Transactions of the 55th North American Wildlife and Natural Resources Conference*. Washington: Wildlife Management Institute.

Westman, W. 1991. "Ecological Restoration Projects: Measuring Their Performance." *The Environmental Professional* 13(3):207–215.

Notes

1. Records in park files indicate the deer herd was very small when the park was established in the 1930s. There may have been deer in locations other than The Hurricane, but sightings were rare. The U.S. National Park Service considered importing deer from Pisgah National Forest, where there was a greater density of native whitetails, but did not go through with the plan.

2. The red wolf has recently been slated for reintroduction in the Great Smoky Mountains National Park, but I doubt it was ever very common in the southern Applachians. The eastern mountain lion may not have been completely extirpated at the time the park was designated, but no major attempt was made to establish its population status in the southern Applachians in the 1930s and 1940s.

3. Information sources for these restoration examples include Bart Johnson who is conducting the mountain avens restoration, Barbara Zoodsma who is studying the manatees in Cumberland Sound, Robert Warren who is supervising the bobcat restoration, and Albert Meier who is working on herbaceous species diversity in old growth forests.

11

Ecosystem Health
and Trophic Flow Networks

ROBERT E. ULANOWICZ

Ecosystem "health" may best be understood in the context of a hypothesized natural tendency for systems to grow and develop. These processes have been quantified using an information-theory measure called the network ascendency. Ascendency by itself is not a proper surrogate for system health, but it does form one index that may be useful in constructing such a description. A healthy system is hypothesized to require both a high diversity of intercompartmental transfers and a high mutual information among them. A newly derived "scope for ascendency" also seems to afford a sensitive assessment of system performance. These measures have been applied to comparative data on undisturbed and impacted tidal marsh creek ecosystems in Crystal River, Florida. The scope index is particularly sensitive to the effects of stresses in the disturbed system. The network data also can serve as input to other analyses that assess the degradation in trophic functioning and energy processing by the disturbed system.

Health is one of those concepts, like information, that is difficult to define in direct terms. Instead, both are delimited by what they are not—that is, by their antonyms, which are more apparent. Uncertainty, for example, is both central to the human condition and easily quantified in probabilistic terms, as Claude Shannon (1948) capably demonstrated. Information, however, can be measured only as a decrease in uncertainty. Similarly, health is most obvious to us by what it is not—the occurrence of disease, trauma, or dysfunction. Webster's New Collegiate Dictionary cites as the most common meaning "freedom from physical disease or pain." Compounding matters for ecologists is the analogy of ecosystem health to the health of the human body or the bodies of organisms. Is the extension of the notion of health to other entities simply a rough metaphor, or is there an ontological basis for speaking of the "health of an ecosystem"? Webster provides a trail of clues beginning with the secondary definition of health as a "flourishing condition" or well-being, as might pertain to the economy of a country. The same dictionary describes "to flourish" as "to reach a height of development or influence." This brings us closer to the realm of ecology, for in that discipline there is a strong tradition of studying a phenomenon known as succession. It appears that pioneer ecosystems colonizing a newly opened area progress through a roughly predictable series of states, culminating in what approximates a climax community, whereafter changes at the level of the entire community become insignificant.

If one can somehow quantify the progress of an ecosystem along its pathway of succession (development), then the avenue toward the measurement of ecosystem health is opened: A healthy ecosystem is one whose trajectory toward the climax is relatively unimpeded and whose configuration is homeostatic to influences that would displace it back to earlier successional stages. I intend to use this definition of ecosystem health as the starting point for quantitative assessment.

Before starting the discussion I should note that the postulate of "disease" as impeded development hardly vanquishes all ambiguities. For example, there is widespread confusion over the relationship between development and evolution. To Eugene Odum (1969) evolution is succession (development) writ large. Eric Schneider (1988) tends to agree with Odum and provides the analogy that succession (development) is to evolution as stud poker is to draw poker. The game remains pretty much the same, but the former version is subject to more constraints. Stanley Salthe (pers. comm.) limits the term "development" to those progressions that unfold toward an endpoint that is knowable in advance. Thus development is unlike the less-certain process of evolution, for which the endpoint remains unclear. The question of endpoint becomes crucial to Crawford Holling (1986), who sees the destruction of the climax community by some surprise agent as one link in the necessary cycle of growth, destruction, and renewal, without which evolution could not occur. Hence the notion of health in the longer evolutionary scheme becomes problematical at best. A healthy ecosystem at one temporal scale would serve to impede evolution over the longer duration.

Such controversies notwithstanding, I believe that focusing on health as related to the progress of succession is a good place to begin quantitative narration. It is entirely possible that a metric tailored to the process of development might still pertain to the process of evolution in the absence of an identifiable endpoint. The betting here is that development and evolution are, as Schneider suggests, manifestations of the same game under different constraints.

ECOSYSTEM DEVELOPMENT

A seminal albeit controversial synopsis of the trends apparent in ecological succession was given by Odum (1969). His list of twenty-four attributes of mature systems is cast at various levels of the hierarchy from the individual to the whole ecosystem and includes such indicators as gross production/community respiration quotient, biochemical diversity, organism size, niche specialization, and "information." Elsewhere (Ulanowicz 1980) I have focused on those attributes that could be identified as properties of quantified networks of trophic interactions. These properties were grouped into four categories: greater species richness; more niche specialization; more developed cycling and feedback; and greater overall activity. On purely phenomenological grounds I then argued that all four of these trends could be encapsulated into a single index: the network ascendency.

Ascendency is the product of two factors, one that gauges the level of system activity and another that captures the degree of trophic organization. The former is taken explicitly from economic input/output theory (see Chapter 12 of this volume). If T_{ij} is the transfer from compartments i to j $(i, j = 1, 2, 3,..., n)$ of some commensurable system product, if exogenous inputs are assumed to derive from a hypothetical zeroth compartment, and if exogenous outputs flow to an imaginary compartment $n + 1$, then

$$T = \sum_{i=0}^{n} \sum_{j=1}^{n+1} T_{ij}$$

is the calculated value of the total system throughput. As T defines the size of the system, any increase in T will be reckoned as growth in the very same sense that economic growth is considered to be any increase in the gross national product.

The organization of the network is slightly more difficult to quantify. Again, the reason is that "organization," like health and information, is best described by what it is not. In fact, it can be argued that there is a one-to-one correspondence between what is perceived as organization and what is measured as information in mathematical terms. If one has no knowledge about how the flows T_{ij} relate to each other, one's uncertainty about the system structure is measured by the Shannon-Weaver index of uncertainty:

$$H = - \sum_{i=0}^{n} \sum_{j=1}^{n+1} (T_{ij}/T) \log(T_{ij}/T)$$

Once one knows how the flows are connected in the network, one's uncertainty is then reduced by an amount known as the "average mutual information":

$$I = \sum_{i=0}^{n} \sum_{j=1}^{n+1} (T_{ij}/T) \log\left[T_{ij}T \middle/ \left(\sum_{k=1}^{n+1} T_{ik}\right)\left(\sum_{m=0}^{n} T_{mj}\right)\right]$$

One can prove for any network of exchanges T_{ij} that $H \geq I \geq 0$. Any increase in the organization I of a system can be taken as development.

The product of T and I is called the system ascendency (A), so named because of the dual meaning the word imparts. In the absence of major perturbation, succession proceeds in the direction of increasing ascendency (Ulanowicz 1986a). (The system "ascends" to more mature configurations.) Looked at differently, a configuration with higher ascendency can dominate over a system with a lower value of A and presumably would displace the latter whenever there is free communication between them.

Although the two factors in the ascendency relate to different concepts, the increase in ascendency can usually be traced to the influence of a unitary agent—autocatalysis, or indirect mutualism. Growth usually does not occur without development and vice versa. I argue that autocatalysis is an agent rather than a mechanism, because it exhibits properties that cannot be traced to the separate behavior of its constituent parts (Ulanowicz 1989).

ASCENDENCY AND SYSTEM HEALTH

If, as has been argued, the natural course of ecological succession is in the direction of increasing ascendency, it would at first seem reasonable to choose ascendency as a surrogate for ecosystem health. While there usually is some correlation between health and ascendency, the correspondence is not exact. Discrepancies between the two can be traced to at least two causes. First, increases in ascendency signify any combination of growth and development. It often occurs that most of the increase in A can be traced to an increase in the scaling factor, T. It also sometimes happens that T can increase at the same time that I decreases, but the increase in T predominates and the resultant product $A = TI$ still increases. This latter situation provides a convenient quantitative definition of the phenomenon of eutrophication (Ulanowicz 1986b). The sudden accessibility to new resources by the producer organisms results in amplified flows among the lower trophic members that overwhelm and often extirpate higher trophic elements. The resultant decrease in species richness and associated transfers causes the mutual information of the flows to drop, but this decrease is more than compensated by the concomitant rise in total system throughput. The system ascendency rises, but the "health" of the system by most standards has definitely suffered.

The size of a system is seen to be only marginally connected with its health. Although size may allow a larger system to displace a smaller configuration, it does not immediately follow that the former is necessarily healthier than the latter. Here an analogy with the human situation is helpful. A 2.3-meter leviathan should have little trouble beating up a 1.2-meter dwarf in a hand-to-hand confrontation (unless, perhaps, the latter's name happens to be David). It does not necessarily follow, however, that the giant is healthier than the smaller person. Thus a descaled version of the ascendency—such as $I = A/T$ (Ulanowicz and Mann 1981) or I/H (Field et al. 1989)—should be more tightly correlated to the developmental stage and thus to ecosystem health.

Even a scale-independent ascendency is not an unerring index of system health. A high value of the mutual information of flows, I, usually indicates narrow trophic specialization. With such "stenotrophy" comes a decline in the number of alternative pathways that could grow to compensate for a disturbance in a major route of nutrient or energy flows. The parallel but secondary routes function as a "strength-in-reserve" with which the system can adapt to unexpected and novel disturbances. Systems that "overdevelop" lose such homeostatic capability and appear "brittle" (Holling 1986) to even slight impacts.

It is not sufficient that I/H be a large fraction in order to conclude that a system is healthy. It is necessary also that both numerator and denominator possess magnitudes as large as possible (Ulanowicz 1986b). The denominator is the diversity of flows, which in turn requires a diversity of compartments for its sustenance. Thus we are led to a heuristic justification for preserving species diversity, an ethic that some suggest needs no rational justification (Sagoff 1988).

Finally, it should be noted that ascendencies are usually calculated from data on static, balanced networks. On the average inputs are balanced against outputs. It may be, however, that either inputs or outputs predominate in a system—that is, the system is growing or shrinking—and the ascendency as formulated is relatively insensitive to which condition might be prevailing at the time of measurement. For this reason one might wish to use an index possessing some antisymmetry that distinguishes inputs from outputs. Winberg (1956), for example, defines the "scope for growth" of a population as the rate by which the inputs to the species exceed its outputs, and Genoni and Pahl-Wostl (1991) have suggested extending this notion to the entire community. Such a straightforward extension, however, fails to incorporate system structure and would likely be too rough a measure to capture even the major factors comprising ecosystem health. In its place I have suggested using the "scope for ascendency," a measure that bears a formal resemblance to the ascendency but distinguishes between inputs and outputs in the system. The equation for the scope is

$$S = \sum_{i=0}^{n} \sum_{j=1}^{n} (T_{ij}/2) \log\left[T_{ij}T\Big/\left(\sum_{k=0}^{n} T_{kj}\right)^2\right] - \sum_{i=1}^{n} \sum_{j=1}^{n+1} (T_{ij}/2) \log\left[T_{ij}T\Big/\left(\sum_{k=1}^{n+1} T_{ik}\right)^2\right]$$

As with the conventional scope, S is usually positive when inflows dominate and can become negative when losses prevail. It is not generally zero at steady state, but more often positive, owing to the limited number of exogenous inputs in relation to the more dispersed nature of the outputs. When the scope is applied to a simplistic mechanical system like the steady-state Atwood's machine,[1] it gives results similar to the power generation function (Odum and Pinkerton 1955; Smith 1976). S is zero for steady-state extremes of component efficiency (0 and 1) but reaches a maximum at some intermediate first-law efficiency. Thus S appears (at this stage of research) to quantify the effectiveness with which the system utilizes the given resources.

APPLICATION TO CURRENT DATA

As the ascendency and associated indices pertain to the state of the entire ecosystem, they therefore require data on all transfers occurring in the community. The collection of such data is usually a laborious task. For this reason, fully quantified networks of ecosystems remain scarce, although recently there has been a spurt of attempts to elaborate flow networks (Pauly, pers. comm.). Despite this encouraging trend, I know of only one set of data (taken over fifteen years ago) systematically assembled to compare the networks of an impacted ecosystem with those of a control counterpart (Homer and Kemp 1975; see also Ulanowicz 1986a). These data were taken from a tidal marsh tributary creek off Crystal River, Florida, and from a similar creek that was subjected to an average 6°C rise in ambient temperature because of exposure to the effluent from an adjacent nuclear power generating station.

The mix of species inhabiting the control and impacted ecosystems closely resembled each other; however, the food web of the disturbed system appeared less connected than that of the control (see Figures 1 and 2). The question here is: Do the ascendency and related variables reflect any degradation in the health of the impacted system? The answer is a qualified yes.

The level of trophic activity in the impacted system was almost 20 percent less than in the control creek. The heated system was obviously operating beyond its optimal temperature. As T scales the ascendency, it showed an almost proportionate decline in magnitude. The unscaled indices all fell in the impacted system, but the decrements were hardly significant. I fell by 1.7 percent, H by 1.1 percent, and I/H by a scant 0.7 percent. The conclusion one draws is that the activity level is sensitive to temperature changes but, at least within this range of temperature shift, the basic structure of the ecosystem remains virtually unchanged. The same response was noted for seasonal changes in the Chesapeake Bay mesohaline ecosystem by Baird and Ulanowicz (1989).

The scope for ascendency fell by over 7 percent in the Crystal River, probably reflecting the impaired ability of the heated system to utilize effectively what resources it does capture. The greater sensitivity of S to structural changes recommends it as a preferred index for gauging system performance.

OTHER INDICATORS OF SYSTEM HEALTH

Ascendency involves an information-theory view of trophic dynamics. There are other ways of looking at the whole system that help to indicate the system's performance, although the relationship of these other portrayals to ecosystem health remains less direct than those of the more fundamental index, the ascendency.

Odum (1969) notes as one of his twenty-four criteria that mineral cycles tend to be more closed in mature ecosystems.

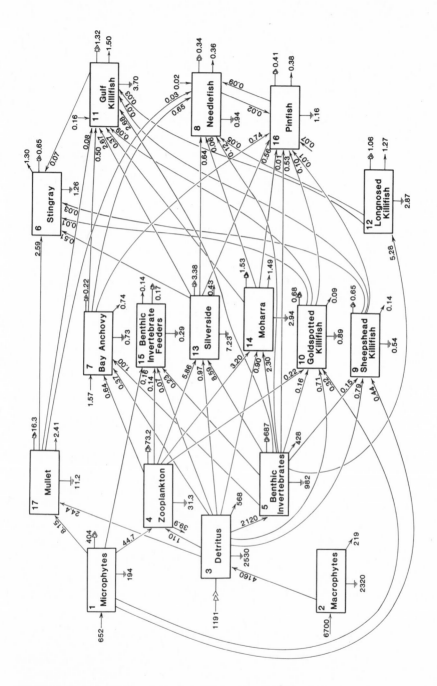

FIGURE 1. *Schematic of Carbon Flows (mg/m²/day) Among the Taxa of a Marsh Creek Ecosystem: Crystal River, Florida. The linked arrows represent returns to the detritus.*

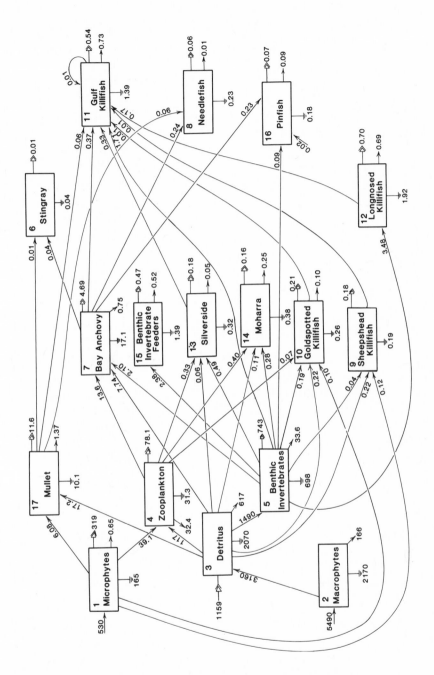

FIGURE 2. *Schematic of Carbon Flows in a Nearby Heated Marsh Creek Ecosystem. This ecosystem is similar to the one depicted in Figure 1, except that it was perturbed by a 6 °C average elevation in temperature due to thermal effluent from a nearby nuclear power generating station.*

John Finn (1976) developed an index that measures the fraction of overall activity devoted to recycling. The implication is that systems with higher values of the Finn index are more mature. While Odum spoke of mineral cycles, Finn and colleagues (Richey et al. 1978) tried to apply the index and the concept to carbon cycling in lakes, only to get very equivocal results. Carbon, however, is tied closely to energy, and the fraction of carbon that is recycled is indicative of the system's inability to utilize fully the energy resources available to it. Hence, a high Finn index for carbon could be a sign of a more stressed community. This appears to be the case with the Crystal River systems, where the Finn index of 7.1 percent in the control creek rises to 9.4 percent in the heated counterpart. Wulff and Ulanowicz (1989) also noted that the Finn index of carbon cycling in the less impacted Baltic ecosystem was 22.8 percent, but rose to 29.7 percent in the more eutrophic Chesapeake community. If anything, the Finn index when applied to carbon flux appears to be a counterindicator of ecosystem health. (Of course, the same index when calculated on the limiting nutrient might yield altogether different results.)

There are other aspects of carbon cycling that bear on the system's performance. If one enumerates the number of cycles in the two creeks (Ulanowicz 1983), for example, one finds a total of 119 simple directed cycles in the control creek, but only forty-six in its impacted counterpart. Furthermore, the cycles in the control creek tend to have more trophic links and involve more of the higher species. There are forty-two cycles of trophic length 4 and twenty-six of length 5 in the control creek, whereas there were only fourteen cycles of length 4 and no cycles of length 5 in the impacted creek.

With respect to food chains, Odum has characterized pioneer systems as resembling linear grazing chains, whereas more mature systems seem more "web like" and involve greater detrital feeding. One can see by inspection of Figures 1 and 2 that the control creek has a higher topological connectivity and thus appears more web like. What Odum fails to mention is that one would also expect more mature or less stressed systems to have longer feeding pathways and more species of higher trophic rank. Stephen Levine (1980) has shown how the average trophic rank of each species may be calculated from a quantified food web. The species with the highest trophic rank in the control creek was the needle-fish (3.46), which tied for the highest rank in the impacted system with the stingray at a value of 3.68. The situation is further clouded by the fact that six species seemed to flourish in the warmer environs. The trophic status of the needlefish, for example, increased from 3.46 in the control community to 3.68 in the impacted one. Likewise, the bay anchovy improved significantly from 2.02 in the undisturbed to 2.67 in the stressed network.

The individual trophic ranks appear to indicate that the heated system is doing better at moving material to higher trophic levels—a conclusion that is avoided once one aggregates both webs into their straight-chain configurations. This mapping involves apportioning the activity of each species to integral, sequential members of a chain according to the fractions that arrive at that compartment over pathways of corresponding integer length (Ulanowicz and Kemp 1979; Ulanowicz in press). The aggregation is done in such a way as to preserve material balance. The results of trophic aggregations are shown in Figure 3.

One notices immediately that the trophic chain representation of the control creek (Figure 3a) is one step longer than that of the impacted system (Figure 3b) and that almost five times as much medium reaches the fourth trophic level in the undisturbed community. The arrow from box I to box II represents pure herbivory (grazing), while that from the box marked D (for detritus) to box II signifies detritivory. Herbivory falls only slightly in the impacted system, whereas the drop in detritivory is more significant. The detritivory/herbivory ratio falls (as Odum predicted) from forty-one in the control system to thirty-six in the stressed, even though a greater proportion of activity in the stressed system is being devoted to recycling. The aggregated trophic chains show unambiguously that the control creek is performing better than the heated ecosystem.

CONCLUSION

Assessing the health of ecosystems requires a pluralistic approach and a number of indicators of system status (Karr 1991; Schaeffer et al. 1988). Ambiguities remain, however, regarding how to relate these indices to the fundamental notion of system development, which is a critical process in defining ecosystem

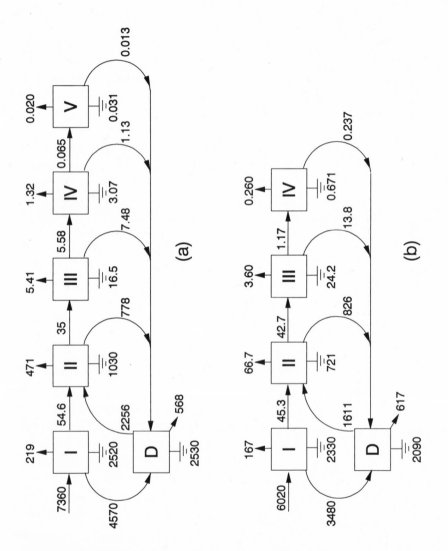

FIGURE 3. *Aggregated Trophic Transfers (mg C/m²/day) Occurring in (a) the Unperturbed Marsh Creek and (b) the Heated Creek.*

health. By contrast, the ascendency measures espoused here require copious data on the workings of the entire system—data that take much effort to assemble. The chief advantage in using these measures, however, lies in their hypothetical connections to what is really meant by healthy system functioning.

As more and more ecosystem networks are elaborated, moreover, the process of estimating flows for entire systems becomes progressively less difficult. Such facilitation stems from a growing pool of expertise, a larger body of published data, and a greater availability of software (such as ECOPATH II by Pauly et al. 1987) that actually assists the investigator in constructing quantified ecosystem networks. Ten years hence, estimating changes in ecosystem flow networks may not appear so formidable a task as it now seems. And with the increasing interest in analyzing flow networks should come the discovery of new relationships between what now appear as ad hoc indices (that is, diversity) and the more fundamental processes of growth and development of ecosystems, which underlie the notion of ecosystem health. Such a justification for the use of population diversity indices in ecosystem management was mentioned earlier, and surely other connections could follow.

Is ecosystem health nothing but a broad metaphor? Some consider the very notion of an ecosystem to be only a mental construct, but they are clearly in the minority. A recent poll of interest by ecologists in various subdisciplines revealed "ecosystems" to be the most popular topic (Waring 1989). As real (and measurable) as ecosystems themselves is the notion that they grow and develop. Systems exhibiting growth and development almost by definition are subject to disruption of these fundamental processes—whence enters the notion of health. No one is suggesting, as some have imputed to Clements and Shelford (1939), that ecosystems are the ontological equivalents of organisms. But from an operational point of view, it makes great sense to attempt to measure a system's growth, development, and health by using indices that can guide (and often temper) our efforts to manage the ecosystems that sustain humankind.

References

Baird, D., and R. E. Ulanowicz. 1989. "The Seasonal Dynamics of the Chesapeake Bay Ecosystem." *Ecological Monographs* 59:329–364.

Clements, F. E. and V. E. Shelford. 1939. *Bio-ecology*. New York: Wiley.

Field, J. G., C. L. Moloney, and C. G. Attwood. 1989. "Network Analysis of Simulated Succession After an Upwelling Event." In F. Wulff, J. G. Field, and K. H. Mann (eds.), *Network Analysis in Marine Ecology*. New York: Springer-Verlag.

Finn, J. T. 1976. "Measures of Ecosystem Structure and Function Derived from Analysis of Flows." *Journal of Theoretical Biology* 56:363–380.

Genoni, G. P., and C. Pahl-Wostl. 1991. "Measurement of Scope for Change in Ascendency for Short-Term Assessment of Community Stress." *Canadian Journal of Fisheries and Aquatic Sciences* 48:968–974.

Holling, C. S. 1986. "The Resilience of Terrestrial Ecosystems: Local Surprise and Global Change." In W. C. Clark and R. E. Munn (eds.), *Sustainable Development of the Biosphere*. Cambridge: Cambridge University Press.

Homer, M., and W. M. Kemp. 1975. "Trophic Exchanges in Two Tidal Marsh Guts." Unpublished manuscript. Gainesville: University of Florida Department of Environmental Engineering.

Karr, J. R. 1991. "Biological Integrity: A Long-Neglected Aspect of Water Resource Management." *Ecological Applications* 1:165–222.

Levine, S. H. 1980. "Several Measures of Trophic Structure Applicable to Complex Food Webs." *Journal of Theoretical Biology* 83:195–207.

Odum, E. P. 1969. "The Strategy of Ecosystem Development." *Science* 164:262–270.

Odum, H. T., and R. C. Pinkerton. 1955. "Time's Speed Regulator: The Optimum Efficiency for Maximum Power Output in Physical and Biological Systems." *American Scientist* 43:331–343.

Pauly, D., M. Soriano, and M. L. Palomares. 1987. "Improved Construction, Parametrization and Interpretation of Steady-State Ecosystem Models." Paper presented at the Ninth Shrimp and Fin Fisheries Management Workshop, 7–9 December 1987, Kuwait.

Richey, J. E. et al. 1978. "Carbon Flow in Four Lake Ecosystems: A Structural Approach. *Science* 202:1183–1186.

Sagoff, M. 1988. "Ethics, Ecology, and the Environment: Integrating Science and the Law." *Tennesse Law Review* 56: 77–229.

Schaeffer, D. J., E. E. Herricks, and H. W. Kerster. 1988. "Ecosystem Health I: Measuring Ecosystem Health." *Environmental Management* 12:445–455.

Schneider, E. D. 1988. "Thermodynamics, Ecological Succession, and Natural Selection: A Common Thread." In B. H. Weber, D. J. Depew, and J. D. Smith (eds.), *Entropy, Information, and Evolution*. Cambridge: MIT Press.

Shannon, C. E. 1948. "A Mathematical Theory of Communication." *Bell System Technical Journal* 27:379–423.

Smith, C. C. 1976. "When and How Much to Reproduce: The Trade-off Between Power and Efficiency." *American Zoologist* 16:763–774.

Ulanowicz, R. E. 1980. "An Hypothesis on the Development of Natural Communities." *Journal of Theoretical Biology* 85:223–245.

―――. 1983. "Identifying the Structure of Cycling in Ecosystems." *Mathematical Bioscience* 65:219–237.

―――. 1986a. *Growth and Development: Ecosystems Phenomenology*. New York: Springer-Verlag.

―――. 1986b. "A Phenomenological Perspective of Ecological Development." In T. M. Poston and R. Purdy (eds.), *Aquatic Toxicology and Environmental Fate*. Vol. 9 ASTM STP 921. Philadelphia: ASTM.

―――. 1989. "A Phenomenology of Evolving Networks." *Systems Research* 6:209–217.

―――. In press. "Ecosystem Trophic Foundations: Lindeman Exonerata." In B. C. Patten and S. Jorgensen (eds.), *Complex Ecology*. Englewood Cliffs: Prentice-Hall.

Ulanowicz, R. E., and W. M. Kemp. 1979. "Toward Canonical Trophic Aggregation." *American Naturalist* 114:871–883.

Ulanowicz, R. E., and K. H. Mann. 1981. "Ecosystems Under Stress." In T. Platt, K. H. Mann, and R. E. Ulanowicz (eds.), *Mathematical Models in Biological Oceanography*. Monographs on Oceanographic Methodology. Paris: UNESCO.

Waring, R. H. 1989. "Ecosystems: Fluxes of Matter and Energy." In J. M. Cherrett (ed.), *Ecological Concepts*. British Ecological Society Symposium 29. Oxford: Blackwell.

Winberg, C. J. 1956. "Rates of Metabolism and Food Requirements of Fishes." *Fisheries Research Board of Canada Translation Series* 194: 202.

Wulff, F., and R. E. Ulanowicz. 1989. "A Comparative Anatomy of the Baltic Sea and Chesapeake Bay Ecosystems." In F. Wulff, J. G. Field and K. H. Mann (eds.), *Network Analysis in Marine Ecology*. New York: Springer-Verlag.

Notes

1. Atwood's machine consists of two unequal weights attached to the ends of a rope that is strung over a pulley. The rope and pulley are usually assumed to be frictionless. When allowed to free-fall, the

heavier weight performs work in raising the lighter one. For the most part, you could consider Atwood's machine as a hypothetical artifact useful in clarifying various issues about work and energy.

12

Measures of Economic and Ecological Health

BRUCE HANNON

To my unreasoned envy, economists have an organizing device for their professional activities upon which they nearly all agree. Most economists agree that the total consumption in an economy should be maximized (Sagoff 1988). Of course, there are restraints on the growth of the consumption rate such as the supply rates of labor and natural resources. So the constrained rate of growth of consumption approaches the "golden" level when full employment is nearly reached, profits are being maximized by resource holders, and the rate of technical progress is steady. The growth rate is thus limited by the birth rate, the rate of discovery of new natural resources, and the rate of technical progress. All three rates are usually taken as given. They have forged this wonderfully consistent story of how the economy works (or should work) over the last century. Economists contend over the means to achieve maximum rates of consumption, but they agree on the general goal. By virtue of the detail behind the total consumption number, economists are able to argue rather consistently about the effects that certain proposed changes will have on this total.

I envy the economists even though I know that their story is incorrect. Their story does not count the environmental and social calamity that our focus on consumption has caused. The costs of this calamity should be counted into the costs of production, say the economists: these costs are simply externalities whose costs must be internalized. But what is the cost of a forest gone or a community disbanded? Who is there to remember and to count these costs and who will "put" them into the cost of someone else's production?

Nevertheless, the focus of the economists on a single goal to which they largely agree should be very instructive to ecologists. That focus has bathed them in public appreciation and professional success. They claim that this goal is the measure of economic health. It is my purpose here to show ecologists how they too can devise a somewhat similar measure of ecosystem health, but one that avoids some of the more blatant pitfalls of the economic analog.

If you ask ecosystem managers about the goal toward which they manage the system under their control, they might say that they strive to stabilize the system with the least human input possible. This is their measure of ecosystem health. They could measure the principal stocks in that ecosystem and consider their management a success if these stocks would cycle through the same levels in years in which the main external inputs (such as rain) are at typical levels and return to these cycles after an unusual year of inputs. Or they might say that they manage their systems to maximize duck or bass or deer or tree production for the special interests supporting their agencies' programs (see Chapter 3 of this volume). Or they might manage their systems to conform to the aesthetic ideals of the foundation that pays their salaries. The system managers might also be ecological historians who wish to restore their systems to their ideas of what these systems were like in prehuman times.

In the broadest sense, these are all measures of ecosystem health. But there are possibly several, if not many, stable states for an ecosystem depending at least on the level and kind of management input (see, for example, May 1977 and Schaeffer 1990.) The management of an ecosystem to maximize the production of a single species is liable to end up sacrificing other species and bring the system to an unstable state. Management for aesthetic reasons is a fickle master as these reasons change when the controlling people die or grow tired of their view. The ecological historians usually cannot determine what the original system was really like and thus sometimes resort to trying to restore the system to the earliest recorded form. If they wish to restore river ecosystems to their original forms but have only anecdotal evidence, for example, they may strive to restore the river to the form they see in the earliest aerial photos (earliest possible: 1930s).

Ecologists need a consistent focus for their idea of ecosystem health, a focus on which most of them come to agree. They must work to achieve consensus on that focus and embed that goal in national and local policymaking. I want to propose a candidate for that goal. It is fashioned after the economist's view of the economic system with an attempt to avoid its pitfalls.

ECONOMIC HEALTH MEASURES

Economists refer to the total consumption of the nation by the general term of "national output" or, in the terms of this essay, "net output." The net output of a modern economy is somewhat arbitrarily but officially defined as the amount of personal and government consumption (cars, food, highways, defense, and so forth), plus the amount of net export of goods and services, plus the amount of new capital formed (investment for expansion and replacement), plus any changes in the inventory of goods, all in a given time period. When taken as an annual sum, this amount is called the gross national product or GNP. The GNP is a flow, not a stock, and it is measured in dollars per year.[1]

All the physical units of things we bought as consumers, the things bought by the government, the exported goods (such as computers) less the imported goods (such as oil), the amount of new capital stock purchased (such as machinery and buildings), the increases in the inventory stocks (such as mined coal), for example, less the decreases of these sorts of stocks (such as unsold cars), are multiplied by their respective prices. This multiplication converts the diverse types of these physical things into a single unit of measure and they can be added together to form the GNP (Peterson 1962).

There is an alternative definition of the same number. The GNP is also the sum of all salaries and wages, plus the sum of all taxes paid, plus the sum of all profits, dividends, and interest payments, plus the total depreciation of capital stocks. When these items are in equivalent monetary units, they too can be added together to give the same dollar value as the GNP. This list is sometimes referred to as the "value added," or in our terms the "net input" to the economy.

So the economists view the economic system as having a set of net inputs and a set of net outputs, the total dollar value of each being equal. They recognize that the value of the input of steel to auto production and the total value of the output of autos are not meaningful measures of the aggregate behavior of the economy. For example, only the value of the steel sold for export or directly to consumers should be counted in the determination of overall economic activity. The dollar value of the steel used by autos will be incorporated into the value of autos. Not all autos are sold to consumers. Some autos are sold to other industries, and the value of those autos is incorporated into the value of the output of the purchasing industries. Therefore, the net output is made up of these net values to the defined targets and the net inputs are the payments required to obtain the services of the agents who are defined as the net output consumers.

If the GNP is high, inflation fears aside, the economy is thought of as being more *healthy* than when the GNP is low or negative. The GNP's value (or changes in its value) is related to the amount of overall employment and total capital expansion. These connections are part of the basis on which economic and social welfare policy decisions are made at the national level.

There is a school of thought in economics which argues that only such macropolicy should be made: the details of the needed changes at the corporate and individual level should be worked out by market interactions. A contending school proposes that more government concern should be directed at the corporate and individual levels. The GNP might be growing while segments of the population are losing their jobs, even though total employment is rising. The economy must also be regulated at the detailed level. The economic positivists contend with the regulators. No doubt both schools have something to contribute, but my point is that the whole question of an appropriate measure of economic health is somewhat unsettled. Both schools would probably argue that GNP growth is at least a necessary if not sufficient condition for a healthy economy, much to the frustration of many environmentalists.

Every item sold in the economy has a distribution of unit values or prices in any given period. The variation of prices depends largely on how tightly the item is defined ("automobiles" vs. "Ford Escort, two door, smallest available engine"). Price changes in an economy are not generally thought of as measures of economic health. In some cases, however, changes in prices that are unrelated to changes in the nature of the product, termed "inflation" or "deflation," are thought of as health measures. Inflation is thought to be controllable from a central authority (the Federal Reserve) by control of the supply of money (the interbank loan rate, the money printing rate, and so on). Price control is largely achieved through indirect control of the discount rate: the interbank premium charged by lenders. Trends in the discount rate, the time value of money, are thought of in some sense as indicative of economic health. Inflation and deflation are thought by economists to be accurately reflected in the discount rate, along with consumer impatience and uncertainty.

Through various optimization models, used in conjunction with the structure and detail mentioned above, economists are able to make predictions about the effect that certain proposed actions will have on the GNP. Most economists who follow such approaches agree on the nature of the approach. Usually, controversy is confined to the accuracy of the parameters used in such an analysis. Economic policy can arise by rough consensus among economists over the nature and impact of various proposals. Such consensus allows them to agree on recommendations to those who manage the political forces in the nation. Such recommendations may not be accepted for a variety of reasons, including the reason that the analysis was not sufficiently inclusive of certain social or ecological factors. The economists then try to correct their analytic approach and evaluate the criticisms. This process has been going on for more than a century, and today economists as a profession have a major influence on policymaking around the world. A basic reason for this success is their agreement on a single unifying measure of economic health, one that is underpinned by an explanatory and predictive body of economic theory which is being constantly applied, challenged, and refined.

Nevertheless, there have been many criticisms of the GNP measure of economic health. (See, for example, American Academy of Political Science 1967; Nordhaus and Tobin 1972; Zolotas 1981;

Daly and Cobb 1989.) I recount here the few that seem to high-light the major problems.

If a homemaker becomes a member of the industrial-commer-cial work force, the new salary becomes part of the GNP. This is an appropriate event if the new job is created in response to a needed expansion of the work force. But suppose that the new salary is spent on home and child care workers, jobs that the for-mer homemaker was doing without pay. These wages then be-come part of the GNP. Both before and after the new employment, the same jobs were being done at the home site, but now the GNP has grown.

Health care expenditures are also included in the GNP. Suppose that an expanded need for health care comes from expo-sure to pollutants that have increased chronic disease. Job atten-dance would drop, but assume that the cost of production would not increase. Assume further that the decreased workplace atten-dance is not severe enough to result in decreased wages. The re-sulting effect is an increase in GNP associated with pollution-bred disease, not a desirable outcome. In fact, we are saying that de-creased human health equates to increased (or unchanged) eco-nomic health. Increases in crime result in increased local law en-forcement expenditures that increase the GNP, provided that the needed increase in taxes does not reduce personal consumption to an offsetting degree. The result may be GNP growth proportional to a growth in the crime rate. Or take this final example: the de-crease in the stocks of finite natural resources due to consumption rates that exceed discovery rates, such as those for oil, iron ore, and fertile soil, are not included in the GNP. Yet these resource inventory changes are very similar to the standard measures that are part of the GNP.

Daly and Cobb (1989) provide the most modern set of correc-tions to the GNP, including income inequities, ownership of capi-tal, and environmental damage to cultural and natural stocks. The problem becomes one of finding an objective measuring scale for these concepts. From the standard GNP definition, one can see that an attempt is made to include the depreciation of capital in the net input and in the net output of the economy. The replace-ment capital is easy to include with relative accuracy in the net output or GNP. It is an unspecified part of the gross investment for the economy. But the actual number included in the net input is the depreciated capital as reported by business. This number includes estimates of financial depreciation (perhaps due to a newly invented machine that produces a less costly product) and dollar values of depreciation that are allowed by specific legislation (to help offset taxes or production costs). These measures may be somewhat unrelated to the physical depreciation, the desired addition to the net input measure.

Is the definition of the GNP a thoroughly good one? No. But the usual conclusion is that the definition could be mended to count a little more here and a little less there, resulting in a meaningful indicator of economic health. Yet there are more general criticisms. The economic basis for the GNP is consumption: more consumption is equated to an increase in economic health. Even if natural resources were infinitely abundant and pollution were not a problem, is there no limit to the possible improvement in economic health? We might try to define an optimal limit to per capita consumption, with restraints on the inequality of distribution across the population. Would there then be no limit on how many people should participate in the economy? Does per capita space become the limiting factor on the growth of total economic health? These questions seem unanswerable because we see our destiny in our own hands, under our own control. We assume that an attitude of continued economic growth is an appropriate policy aim and, because we are in control of Spaceship Earth, we will know when to shift to a more appropriate goal. But there clearly are such things as finite natural resources, and limited and unjust distribution of income and space—things apparently, to some extent, beyond our control. Can we look to the ecologists to find more appropriate measures of economic health?

ECOSYSTEM HEALTH MEASURES

The control of the economy is a subjective enterprise because the generally accepted measures of economic health are ultimately subjective. When we look at another system, the ecosystem, one that we tend to observe from outside, can we in fact measure that system's health less subjectively? The measures of the economic system cited earlier can be used as a guide. Imperfect as it seems, the economic system measures are the result of a great deal of effort over centuries by some of the best minds in economics. Perhaps a measure of ecosystem health can be found that, by re-

verse analogy, will reduce the subjectivity of the measures of economic health. As outside observers of the ecosystem, we might
assign more objectivity to any conclusions we reach from our observations.

A more difficult problem lies with the ecologists. To gain consensus on a measure of ecological health, they must assume that
an ecosystem somehow behaves as though it has a goal or an optimal state. At some reasonable level of aggregation, the ecosystem
has what seems to be a preferred state, just as the human body
seems to have. Disturbances ("disease") are short term perturbations that the system tries to eliminate and then return to its
optimal state. This assumption alone would require a very great
leap—too subjective a move for many ecologists. But I see no alternative to it. If we do not accept the idea of a goal or accept at
least the possibility, we are open to the accusation that ecologists
are arrogant with regard to nature and can manipulate its processes for the will of humankind without risk of ultimate failure.
Dismissal of the possibility of the actual existence of such a goal is
unscientific. We must assume first that there is some internal goal
for the development of an ecosystem, some description of optimal
health. Through a combination of theory and experimentation on
specific ecosystems, we should be able to discern if the growth of
an ecosystem conforms to predictions based on hypotheses about
the goals of growth. Eventually we should be able to describe the
existence and nature of any internal goals.[2] If we can describe
those goals, we then have a basis that most ecologists would likely
agree is the management basis for ecosystems. In an effort to capture the general system experience of economists and demonstrate
the utility of a single ecosystem goal or health measure, I now
propose the ecological equivalent of the GNP. (Similar measures
proposed by others are discussed later.)

The components of an ecosystem—producers, herbivores, carnivores, decomposers, and the like—can be arranged into an eco-
nomic-like accounting framework (Hannon, Costanza, and
Ulanowicz 1991) with interconnecting flows: herbivores eat producers, carnivores eat herbivores. All components give off heat in
the process of using their inputs, and the producers absorb sunlight (among other abiotic substances). A net input and a net output can be defined for an ecosystem. Based on these net flows, a
set of system-wide indicators or ecosystem "prices" can be derived
(Hannon, Costanza, and Herendeen 1986). The concept of dis-

count rates can also be defined for the ecosystem (Hannon 1984). The end result of such an approach is a view of the ecological system that is parallel to the economist's view of the economic system. Such congruency is useful if the two systems are ever going to be meaningfully combined into the same framework.

The first step in deriving a measure of ecosystem health is to define precisely what is meant by net ecosystem output. These net outputs should include the net exports of biomass and any abiotic substances, and the gross additions to stocks. But what is the ecological parallel to personal and government consumption? Return to the economic framework for a moment. Imagine an isolated economy at steady state. The imports and exports are zero and gross capital formation is entirely used to replace worn-out capital. Suppose further that inventory changes are negligible. The only contributions to the GNP would then be the consumption by people and government and the rate of capital replacement. Some have argued that consumers and government are simply producing-consuming sectors in the production economy, similar to any other industrial-commercial sector (Costanza 1980; Costanza and Herendeen 1984). Consumers supply labor and use housing, food, clothing, transportation, education, and so forth to do so. Governments consume in order to direct and control. Technically, these two sectors can be considered part of the industrial exchange pattern and removed from direct involvement in the GNP measure. In our closed steady-state economy, the only GNP measure would then be the capital replacement rate.

With the consumption sectors removed from the net output, some of the earlier criticisms of the GNP measure would vanish. Crime, health care, and homemaker's labor are considered costs of production of household or government services. By analogy, this reduced GNP definition can be matched in the net output definition for an ecosystem. We have no reason to put any particular species or trophic level in the special position of being a component of net output. To the now "objective" human observer, no species or trophic level seems more important than another. The ecosystem net output for a given period can therefore be defined as the net exports and the gross changes in each of the stocks. For the closed steady-state ecosystem, the only net output is the replacement rate of the stocks, making up the losses caused by metabolism and decomposition. This replacement rate is exactly analogous to the capital replacement rate used by economists.

Analogous to the net inputs to the economy (wages, taxes, profits, and so on) is a list of net inputs to the ecosystem. Think of these net ecosystem inputs as necessary for the production of the ecosystem's net output. Among the various forms of net input to a particular ecosystem, such as nitrogen, phosphorus, or sunlight, one of them is generally limiting the size of the system. Assume that the limiting net input is sunlight. With the list of absorbed sunlight quantities, a set of ecosystem sunlight "prices" can be calculated. This set of prices can be multiplied with the associated terms in the net output. The now commensurate result can be summed to give an evaluated net output of the ecosystem under study. This evaluated output is analogous to the redefined GNP: it captures all the activities in the ecosystem, in a weighted manner, and gathers them into a single number. That number—the gross ecosystem product (GEP)—is a viable candidate for aggregate ecosystem health. But the GEP is not a wholly adequate measure, for the GEP, like the GNP, cannot increase forever. We will return to this problem later.

As an application of this ecosystem health measure, the ecosystem input (GEP) was evaluated for two tidal marshes. One of the marshes existed under natural conditions; the other was elevated on average $6^{\circ}C$ by the waste heat from an electricity generating station (Hannon 1985). The net output was evaluated for seventeen biotic sectors of both marshes in terms of absorbed sunlight energy. The evaluated net output of the heated marsh was 33 percent lower than the same measure for the natural marsh. This comparison of systems, however, is the only one of its kind. The comparison is based on a detailed exchange between all the components in the companion marsh ecosystems, a rare combination of ecoscience and research support. This study demonstrates the manner in which the GEP indicates ecosystem health: both marshes are probably close to the steady-state, but the unheated one has a higher GEP. This comparison also reiterates the fact that my favorite measure of ecosystem and economic health is based directly on system flows, not on stocks.

We turn now to other measures of ecosystem health. David J. Rapport (1984) gives five indicators of ecosystem stress: rapid nutrient loss, decline in primary productivity, diminishing size distribution and diversity, and system retrogression (toward early stages). He notes that, in general, such system-wide indicators appear late in the breakdown process and that species-directed fieldwork is needed to provide early warnings (as in the case of Herring Gulls and DDT). These indicators have the advantage that data for them can be easily obtained, but they lack the utility of a single quantitative measure. To varying degrees, the evaluated net output measure captures the ecosystem changes of which Rapport speaks.

James R. Karr (1991) has defined the Index of Biotic Integrity (IBI) which is the sum of the "scores" given each of 12 measurable attributes for an ecosystem. The independence of the various attributes and the completeness of the set of twelve is not seriously questioned. The relative weightings given the attributes are somewhat arbitrary, however, epecially if some of them are interrelated. Such measures, based on good intention and sound ecological judgment, may have as their chief asset the fact that they can be easily measured. But the same fate awaits them as has undercut the additional measures proposed for the GNP—that is the quantitative arbitrariness of each attribute sinks it into controversy.

The ecological health measures of Rapport and Karr are useful in rating or comparing ecosystems. The rated set of ecosystems can augment environmental policy by showing which streams, for example, should be left alone and which might require pollution control. This *a posteriori* form of reasoning, however, does not allow one to indicate the probable outcome of a proposed change to an ecosystem. For such predictive purposes, a model of the ecosystem is needed, preferably dynamic and most likely nonlinear. The first predictive modeling steps, however, can be taken with comparative statics (Hannon and Joiris 1989). The economists have struggled for a long time to provide such a model for economic forecasting. Ecologists are still centered on the idea of an empirical set of ecosystem indicators.

CONCLUSION

Rapport's view of size diminishment and reversion to a juvenile state are effects that are also captured in the time-value measures of a dynamic system. Ulanowicz and Hannon (1987) point out that the "discount" or "interest" rate of an ecological system captures both the average metabolic rate and the average specific energy content for the living members of the system. Hannon (1990) argues that this rate is both the average and marginal rate of specific metabolism[3] for the living elements in a steady-state ecosystem. He further argues that the steady-state evaluation would reveal the lowest discount rate ever experienced in the development of that ecosystem. It is well known that larger species have smaller specific metabolic rates compared to the physically smaller species. One very general way that the average metabolic rate of an ecosystem becomes smaller is for large individuals to displace the smaller ones.

Climax species seem to have been selected as the least mainte-nance-requiring in order to store the most order in the climax ecosystem. If this size shift is assumed to be a goal of the system, an additional constraint on the gross ecosystem product measure is required. The GEP should increase continually but less and less rapidly until it becomes steady. At that point, the system has the largest GEP and the smallest discount rate possible. The combined requirement ensures that the low entropy retained in the system is maximized. If the principal low-entropy input to the ecosystem is sunlight, the optimal system would be storing the most converted sunlight possible.

Thus we have two indicators of ecosystem health: increasing net output of the ecosystem (GEP), albeit at slower and slower rates, and a decreasing aggregate discount rate. We have no good ecosystem measure of the discount rate, however, except perhaps at the climax or steady state. But some constraint on the GEP is needed for it clearly cannot increase forever. The decreasing dis-count rate concept does tell us that the maturing ecosystem will need increasing amounts of biomass whose unit or specific metabolism is increasingly lower. This is usually the case with the introduction of species with larger members, as Rapport has noted. Therefore, if the larger species are disappearing from an ecosystem, its measured discount rate would be rising.

The flow of low entropy from the external environment puts a cap on the size and therefore the GEP for the aggregate ecosys-tem.[4] How does such a view translate back to evaluation of the economic system? Were it not for our discoveries of metallic ores and the stored energy in fossil and nuclear fuels, identical rules would apply for the optimal growth and ultimate climax level of the economic system. But profitable access to vast quantities of stored low entropy allows the economic system to grow well be-yond its agrarian bounds. The GNP, by the standard or by any of the modified measures, is much larger than it would have been without the discovery and use of these vast stores.[5] The average GNP per capita and the total population are both likely to exceed the agrarian bounds. The ecosystem GEP is diminished by the GNP. The GNP and the GEP are competing measures until we can somehow learn to incorporate the economy into the ecological framework.

Are we faced with two indicators of economic health? One indicator is based on continued GNP growth. To use this measure, the known low-entropy stores of the earth must be increased. The net growth in the stores of proved low-entropy energy reserves (discovery rate greater than consumption rate) is then the best short-term indicator of economic health. In the long run, these reserves will be reduced to levels that are not economical to extract. Then the economy will tend toward the agrarian bound—but from above, rather than from below. We will be required to reduce our GNP per capita and probably our population to comply with the solar constraint. Such a change is unprecedented in recorded world history, except perhaps for Ireland. The change would be accompanied by a trauma proportional to the difference between the present level of population and GNP, the ones set by the solar constraint and the time available for the transition. What are the GNP and population level for the agrarian world? The difference between these calculated limits and the current values is a second long-term economic measure of health.

We may rest content with the idea of the incompatibility of the economic and the ecological systems and their GNP and GEP "health" measures described here. The economic system's description with its prices and production measures seems pertinent to human purposes but ecological processes may be proceeding toward a different and as yet unknown goal.

We may want to push the concept further and try to combine these two net output measures in a single indicator of system health. It has been suggested (Isard 1968; Daly 1968; Hannon et al. 1991) that it is possible to place all economic and ecological transactions in the same matrix or accounting framework. Such a framework would contain the intereconomic and the interecological transactions. It would also have to contain all the important transactions between the economic and ecological systems. The net inputs and outputs of the combined system must be in consistent units of measure. This requirement simply means that the units of measure for each row of the transactions (from the row of steel production to the column of auto production, for example) must be the same (say dollars). The requirement of identical units of measure for all the exchanges in a given row means that the inputs to the ecosystem of pollution by the steel industry (say SO_2) must be collected in a special column, the row version of which is the tons of SO_2 absorbed by the various parts of the ecosystem. Likewise, the inputs to the economic system must be measured in consistent units. The wood, for example, harvested from the nation's forests and used in the various manufacturing industries and in the ecological system could be measured in tons of carbon.

To make the net output of the combined system sum to a single meaningful number, a set of system-wide prices must be devised. As we saw with the ecological system, to calculate this set of prices we must first decide on the limiting net input to the combined system. Is it labor or energy, for example? If such a critical input can be identified, the system-wide prices can be calculated and multiplied times the net outputs, changing each of the net output entries into commensurate units. The sum of these converted net outputs is the evaluated net output of the combined system. It is a single measure of the "health" of the combined system. It is a measure that would show the net effect of pure economic growth combined with ecological decline: the combined measure may reveal a decline while the economic measure shows an increase.

One of the problems with this combined net output approach is data. We simply do not know all the ways in which the economy has an impact on the ecosystem (such as the effect of organic chemical air releases on fish) nor all the ways in which the ecosystem provides free services to the economy (such as wind dispersed pollution). Thus a combined calculation would probably never be complete. But any step toward combining the two systems, such that the net outputs can be joined in a common measure of system health, is a step in the right direction.

References

American Academy of Political Science. 1967. "Social Goals and Indicators for American Society." *Annals*, Vols. 371, 373.

Costanza, R. 1980. "Embodied Energy and Economic Valuation." *Science* 210:1219–1224.

Costanza, R., and R. Herendeen. 1984. "Embodied Energy and Economic Value in the United States Economy: 1963, 1967 and 1972." *Resources and Energy* 6:129–164.

Daly, H. 1968. "Economics as a Life Science. " *Journal of Political Economy* 76:392–401.

Daly, H., and J. Cobb. 1989. *For the Common Good.* Boston: Beacon Press.

Hannon, B. 1984. "Discounting in Ecosystems." In A. M. Jansson (ed.), *Integration of Economy and Ecology—an Outlook of the Eighties.* Stockholm: University of Stockholm.

———. 1985. "Ecosystem Flow Analysis." *Canadian Bulletin of Fisheries and Aquatic Sciences* 213: 97-118.

———. 1990. "Biological Time Value." *Mathematical Bioscience.* 100:115–140.

Hannon, B., and C. Joiris. 1989. "A Seasonal Analysis of the Southern North Sea Ecosystem." *Ecology* 70(6):1916-1934.

Hannon, B., R. Costanza, and R. Ulanowicz. 1991. "A General Accounting Framework for Ecological Systems: A Functional Taxonomy for Connectivist Ecology." *Theoretical Population Biology* 40:78–104..

Hannon, B., R. Costanza, and R. Herendeen. 1986. "Measures of Cost and Value in Ecosystems." *Journal of Environmental Economics and Management* 13:391–401.

Isard, W. 1968. "Some Notes on the Linkage of the Ecologic and Economic System." *Regional Science Association Papers* 22:85–96.

Karr, J. 1991. "Biological Integrity: A Long-Neglected Aspect of Water Resource Management." *Ecological Applications* 1(1):66–84.

May, R. 1977. "Thresholds and Breakpoints in Ecosystems with a Multiplicity of Stable States." *Nature* 269:471–477.

Nordhaus, W., and J. Tobin. 1972. "Is Growth Obsolete?" In *Economic Growth.* National Bureau of Economic Research General Series, no. 96E, New York: Columbia University Press.

Peterson, W. 1962. *Income, Employment and Economic Growth.* New York: Norton.

Rapport, D. 1984. "The Interface of Economics and Ecology." In A. M. Jansson (ed.), *Integration of Economy and Ecology—an Outlook for the Eighties.* Stockholm: University of Stockholm.

Sagoff, M. 1988. *The Economy of the Earth.* New York: Cambridge University Press.

Schaeffer, M. 1990. "Multiplicity of Stable States in Fresh Water Systems." *Hydrobiologica* 200–201:475–486.

Ulanowicz, R., and B. Hannon. 1987. "Life and the Production of Entropy." Proceedings of the Royal Society of London., B 232:181–192.

Zolotas, X. 1981. *Economic Growth and Declining Social Welfare.* New York: New York University Press.

Notes

1. We could consider the value of a country's wealth as a measure of its economic health. Wealth is a stock, however, and terribly hard to measure accurately. Since the physical basis for wealth is largely unmarketed on a regular basis, only vague evaluation can be made.
2. I do not see such an existence assumption as a problem. Children learn the complex procedure of "pumping" their playground swing while the theory to demonstrate such behavior requires Newtonian mechanics and an understanding of energy principles. Engineers long ago proved that complex beam deflection can be exactly predicted from a theory which says that the beam deflects in such a way as to minimize its internal strain energy—and does so without a single course in differential equations or optimization techniques!
3. The rate of respiration per unit of stock respiring; the units are one/time.
4. Perhaps sunlight absorption is the limiting factor or perhaps sunlight embodied in water is limiting, as in desert ecosystems. Or maybe the limit is set by the rate of phosphorus extraction as is the case with certain aquatic ecosystems.
5. Gaiaists alert! What purpose would Gaia have in creating the vast organic fossil stores?

13

Ecological Integrity: Protecting Earth's Life Support Systems

JAMES R. KARR

The Science Advisory Board of the U. S. Environmental Protection Agency (SAB 1990) recently urged the EPA to "attach as much importance to reducing ecological risk as it does to reducing human health risk." Implicitly, the SAB suggestion recognizes widespread evidence that past regulatory programs have not protected the nation from continuing environmental degradation and challenges the long-held assumption that the agency's mandate could be accomplished using a human-health paradigm. Moreover, it establishes an ecological health goal based on a human-health metaphor and conjures up an image of a cadre of health professionals attempting to cure the diseases that afflict earth's life support system.

Many people have advocated greater emphasis on ecological risk because of environmental deterioration—manifest in over-harvest of forest and marine resources, degradation of soil and water resources, the threat of global climate change, and reductions in biological diversity, including the threat to cultural identities and self-rule for many human societies. But little has been done to implement the change until now. Why has it taken so long to recognize that cumulative and largely irreversible effects of human carelessness threaten the natural habitats of the earth, including the stability of human habitat?

For thousands of years, the primary interactions between humans and their environment occurred on relatively small spatial and temporal scales. Disease, acts of predation, limited food supplies, and accidents were the major threats to the health or survival of individuals or family groups. Because these threats were concentrated on the individual, the practice of medicine developed with an emphasis on efforts to cure diseased persons. Even population-level threats such as bubonic plague in the fourteenth century did not divert attention from the health and survival of the individual. Improved medical practice and advanced technologies (via the agricultural and industrial revolutions) enhanced the quality of human life. But these technologies also brought environmental costs: negative effects that degraded the quality of human life. Chemical contamination of the environment is an example of a negative effect of technology; contaminants may be materials present at concentrations above natural background levels (such as heavy metals) or chemicals synthesized by humans.

Typically, environmental costs were not spread evenly across all members of society. Disenfranchised groups (children, the poor, racial and ethnic minorities, underdeveloped regions exploited by extractive economies) were at greatest risk. As the chemical threat became more widespread and a social conscience began to grow, the focus of the health professions expanded from diseases caused by organisms to include diseases caused by exposure to a diversity of contaminants. The growing need to protect environmental health (usually narrowly defined as human environment and human health) prompted universities, government agencies, and public service organizations to focus on human health at the population level. Prevention programs and population risk assessments expanded, often stimulating the development of public health schools and environmental health departments. Diseases and chemical contamination remained the chief areas of emphasis. The concept of ecological risk did not develop in the medical community because its philosophy and technology dissociated human welfare from its dependence on earth's life support systems.

During the same period, alternative environmental ethics developed (Callicott 1991). Gifford Pinchot's consumption-oriented "resource conservation ethic" called for harvests to provide the greatest good for the greatest number of people for the longest time. The central theme was the utilitarian value of natural resources for harvest by humans. In contrast, the "preservation ethic" of John Muir suggested that spiritual needs should take precedence over material needs. Muir advocated designation of wilderness areas to fulfill those spiritual needs. These approaches to environmental ethics established an artificial and counterproductive dichotomy between maximum extraction and maximum protection—a dichotomy which persists today in the national debate that pits advocates of spotted owl protection against those who would harvest remaining old-growth forest.

As the Muir and Pinchot debate developed, an alternative approach unfolded in the writings of Aldo Leopold. While Leopold used the by now well-known "land ethic" phrase, his intentions were broader when he said: "A thing is right when it tends to preserve the integrity, stability, and beauty of the biotic community. It is wrong when it tends to do otherwise" (Leopold 1949:224–225). Today, we see the threat to the productivity of land as but one example of a wider biotic impoverishment: the systematic reduction in the capacity of earth to support living systems (Woodwell 1990). The conservation biology movement overlaps the philosophy of Leopold. It began early in this century with a narrow focus on charismatic megafauna and species subject to sport and commercial harvest. More recently it has captured the attention of scientists, politicians, and the public with a general call for protection of biodiversity.

To avoid the terrestrial context of "land ethic" and to provide a clearer mandate, I prefer to use the term "ecological (or biological) integrity ethic" (Karr 1992). An "ecological integrity ethic" is convergent with several trends in environmental philosophy and ethics, the resource use and preservation ethics of Pinchot and Muir, and the broader perspective of the conservation biology movement. Such an ethic also provides a new perspective for those concerned about human health because it explicitly recognizes the dependence of human society on life support systems. One could argue that society does not depend on all species any more than individuals depend on all of their liver cells or on every member of society. Individuals do, however, depend on the functioning of their liver and on the collective wisdom and actions of society. We also value many of the individuals in society for diverse values that cannot be expressed in economic terms (such as spouses, children, distant relatives, friends, or celebrities like Magic Johnson). Similarly, many choose to support the protection of species (through the Endangered Species Act, for example) because of values that similarly cannot be expressed in economic terms. I consider ecological health or integrity to be an umbrella goal, the maintenance of which motivates all environmental legislation. At present there is no broad act covering ecological or biological integrity, although the Clean Water Act comes close as do efforts to pass a biodiversity bill (Studds 1991).

In summary, the growth of human populations and the expanding influence of their technology produce environmental degradation at levels that the conventional medical community or the environmental and public health communities have not dealt with—and perhaps cannot deal with. Our ability to protect ecological health will be determined by our ability to recognize the common underpinning of the biodiversity (or, more aptly, biotic impoverishment) crisis, the proliferation of human disease (physical and mental), and many forms of social injustice. To do that, we need a new generation of "physicians of the environment"—men and women with ecological backgrounds who see the biota of earth as a complex and essential support system for human society. Like medical pathologists, environmental pathologists must have knowledge of ecologically healthy as well as degraded biological communities if their prescriptions are to be taken seriously. Our normal solutions—medicine, technology, and wilderness preservation—cannot be viewed as panaceas to resolve the ills of modern society; they are necessary but not sufficient. Technological solutions developed without sufficient knowledge of their secondary and tertiary effects must be replaced by solutions that reflect comprehension of complex biological systems.

HEALTH VS. INTEGRITY

Health and integrity have multiple meanings. "Health" may be used in a human context (the condition of being sound in body and mind or spirit; freedom from physical disease or pain) or an economic context (a flourishing condition, such as the economic health of a country). "Integrity" implies an unimpaired condition or the quality or state of being complete or undivided. As metaphors for guiding our general goal of protecting biological systems, for avoiding biotic impoverishment in all its dimensions, both are useful although neither is perfect.

Some have criticized the health metaphor because, unlike human health, the ecological health expectation changes. But norms for human health (oxygen consumption and metabolism with altitude, regional differences in cholesterol levels, tolerance to cold) may change just as norms for ecological systems vary with climate, elevation, or geological substrate. Glenn W. Suter, II (1991), for example, has argued that a primary weakness of the health metaphor is that the medical community has historically concentrated on curing disease while efforts to protect ecological health are directed toward prevention. Moreover, he notes that no catalog of ecosystem diseases exists. I contend that these criticisms reflect weaknesses in the medical approach to human health. Indeed, with greater frequency the medical community is targeting prevention as a goal and placing emphasis on routine preventive care (for example, the growth of HMOs) and, as the AIDS crisis demonstrates, our catalog of human diseases is far from complete. Recognition of the need to prevent disease occupies a growing share of the efforts of medical and public health communities. Similarly, efforts to prevent environmental degradation are becoming more widespread, primarily because of the high, long-term costs and other negative consequences of deferred environmental regulations (Costle 1979).

ECOLOGICAL GOALS

Scholars concerned about ecological integrity face a serious dilemma. Because ecological knowledge is incomplete, prediction of the consequences of societal action is an inexact science. With a desire to appear objective and reasonable, ecologists accept the proposition that "tight causal proof is the only basis for instituting controls" that will protect ecological health (Woodwell 1989:14). Woodwell concludes, and I agree, that a hesitancy to accept a preponderance of evidence (as opposed to causal proof) and a willingness to compromise stringent requirements for avoiding biotic impoverishment is the "epitome of unreasonableness." He continues: "Indulgence in the hyperobjectivity now being pushed on science lends support to avarice [while it] destroys the credibility of science and scientists as a source of common sense" (p. 15).

The limited ability of ecologists to predict events and ecological patterns is not unusual among the sciences used to guide societal actions. The National Bureau of Economics, using the most powerful econometric models and with a massive data base, failed in 80 percent of its predictions of even the sign of change of the gross national product (*Business Week*, 27 July 1982), and the Heisenberg Principle and quantum mechanics of physics are probabilistic rather than absolute predictors.

The obvious weaknesses of a health metaphor or the ambiguous integrity goal, when combined with inadequate information and the reluctance of scientists to contribute to the environmental policy dialogue, limit the use of ecological knowledge. Nevertheless, considerable energy has been expended exploring the philosophical and historical underpinnings of the ecological integrity or health ethic (as in this volume). While I believe that scholarly debate on the intellectual, philosophical, and practical merits of the health metaphor is essential, I suggest that the outcome of these discussions will not change the reality of biotic impoverishment and the need to find creative ways to assess the status of the diverse dimensions of life support systems from local to global scales. Because many dimensions of ecological systems are valued by society, the key component of an effort to assess health or integrity must be a pluralistic foundation in ecological principles and theory (Karr et al. 1986; Karr 1991; Schoener 1986; Norton 1991).

In Chapter 5, Talbot Page warns us against village thinking. Just as residents of neighboring villages may distrust one another, neighboring disciplines may dismiss the proposals of other disciplines as "nonintuitive, ad hoc, and not compelling." An added danger is that the "villages" of ecology will battle while biotic impoverishment continues unabated, making ecologists modern Neros fiddling while Rome burns. Future generations will judge us harshly if we continue to do battle against one another rather than joining forces to use the best available information to reverse the trend. Perhaps the strongest argument for joining forces is the complexity of biological systems. This complexity demands an approach that is as redundant as the systems themselves.

ECOLOGICAL INTEGRITY

A biological system, whether individual or ecological, can be considered healthy when its inherent potential is realized, its condition is stable, its capacity for self-repair when perturbed is preserved, and minimal external support for management is needed (Karr et al. 1986). This definition is consistent with that proposed in Chapter 1 of this volume. Stability is not the stability implied in the Clementsian view of the community; rather, it is compatible with the widespread recognition that biological systems are metastable (Botkin 1990).

Systematic assessments of the status of ecological resources (ecological integrity) must first and foremost be based in biology because environmental degradation inevitably is tied to a failure of the system to support biological processes at a desired level. Thus, direct assessment of that biology is essential to an evaluation program. Further, the most appropriate measures of health or integrity neither have the same value in all places nor do they consistently maximize any single biological attribute (such as biodiversity or primary or secondary production). The number and identity of species, local productivity, and other attributes vary from location to location as a function of environmental conditions (climate, topography) and site history (physical and biological).

Implicit in the phrase "biotic impoverishment" is the notion of a biological benchmark or reference condition against which a site should be evaluated. That benchmark defines the characteristic biota with little or no influence from modern human society. A common question is: Do actions of modern society alter the long-term sustainability of the endemic biota? If human actions alter that sustainability, society is faced with the need to decide whether the alteration is acceptable. At the local level we make these decisions all the time—when we modify vegetative cover to build a house, construct a road, or till the land or modify a river to make it navigable. Local alterations of biological integrity have long been acceptable. But we now find that as they accumulate across the landscape, the sum of these alterations to biological integrity exceeds the sum of the individual actions.

Ecologists and society face a critical challenge in the next few years. No generally accepted approach is available to measure these local and regional impacts; we do not have an ecological indicator for use on the evening news that is comparable to the GNP (Meadows 1990). The development of indicators is, arguably, the most important step needed to mobilize social support for reversal of the trend toward biotic impoverishment. Much of this volume is devoted to documenting on technical or philosophical grounds the need for such assessment methods. What are the critical components of the biological systems that must be protected? Decisions about these components are difficult because very divergent views are held by ecologists. There are widely known empirical relationships between human action and assemblage attributes that can be used to guide society's decisions today. Many ecological concepts (resilience, resistance, connectance, ascendancy) are also thought to be important indicators of a system's condition, although practical application of these concepts is limited because of the difficulty encountered in defining them operationally and measuring them in real-world applications.

The multivariate nature of complex biological systems requires that evaluations be based on a number of relevant biological attributes. A comprehensive effort to protect ecosystem health or ecological integrity should be designed to protect the features that are familiar to our cultures—features that support essential ecological services and long-standing earth processes. An evaluation should include attributes from a hierarchy of levels in ecology from individual health to attributes of constituent populations or assemblages of species (Karr 1991). They should be based on empirical and theoretical understanding of the properties of biological systems, especially their properties when subjected to human impacts. As continued use of aspirin demonstrates, theoretical understanding is not essential.

Key components of biological integrity have been measured before. Toxicologists recognize the importance of individual health in evaluating the extent to which human actions have degraded ecological integrity. Ecologists measure species richness or the relative abundance of species to evaluate ecological conditions. Fish biologists use measures of population age structure to assess the health or vitality of target populations. While none of these attributes alone is broad enough, collectively they can be very diagnostic of local conditions. Although this collective approach should be comprehensive as it represents the many important components of ecological systems, it need not include all possible components (Karr 1991).

An array of attributes used to assess ecological integrity should encompass both the elements and processes of biological systems (Karr 1992). The species of plants and animals are the most obvious of the elements. Other levels of concern are the genetic diversity within species and the presence of an appropriate community and landscape complex. Processes (or the functional relationships) also span the hierarchy from individuals (metabolism) to populations (reproduction, recruitment, dispersal) to assemblages (nutrient cycling, interspecific interactions, energy flow). The main issue is not that some array of attributes taken from this complex is essential. Rather, efforts to assess ecological integrity are more likely to detect degradation in all its potential forms if they are conceptually diverse. Selection of attributes must balance the need for information with the cost and time involved in collecting that information. Direct measurements are best. But when direct measurements are prohibitively expensive or in other ways difficult to implement indirect measurements may be necessary. Measurement of primary production may not be practical, for example, but use of an assemblage's trophic structure may provide an acceptable first approximation of system trophic dynamics.

BIOLOGICAL INTEGRITY AND WATER RESOURCES

Twenty years ago the Water Quality Act Amendments of 1972 (PL 92- 500) first used the phrase "biological integrity" when it called for the restoration and maintenance of "the chemical, physical, and biological integrity of the nation's waters." The sum of chemical, physical, and biological integrity is ecological integrity. Progress toward accomplishing this mandate was slowed because the dominant approach of water resource managers (Karr and Dudley 1981) was based on the chemical and disease emphasis of human health agencies and the narrow chemical, point-source, and technology-based focus of regulatory agencies. The fallacy of that approach and the underlying assumption upon which it was based is now obvious as the quality of water resources continues a downward spiral—witness, for example, the status of the Great Lakes (Edwards and Regier 1990), the condition of the fishery of the Illinois River (Karr et al. 1985), and the number of threatened fish stocks in North America (Williams et al. 1989; Nehlsen et al.

1991). Simply put, the biological integrity of these resources continues to decline because chemical contamination has been the sole focus among a myriad of human influences that have degraded the water resource (Karr 1991). Other consequences of human actions are degradation in physical habitat, altered flow regimes, changes in the organic input to waterways, and changes in biotic interactions such as predation, competition, disease, harvest, or introduction of exotics.

Programs to protect the quality of water resources (a broader concept than water quality) are changing, however, especially with widespread adoption of biological criteria for assessment of the status of water resources (EPA 1990). Even the definition of pollution was expanded in the 1987 Clean Water Act when it was defined as human-induced alteration of the water resource. This definition does not focus on chemical contaminants; rather, a broad perspective is being adopted. Under this definition, humans may degrade or pollute by withdrawing water for irrigation, by overharvesting fish populations, or introducing exotic species. The transition to a more informed management program came in the 1980s with the implementation of integrative biological approaches.

Concrete examples of a rapidly evolving social perspective include the development of National Program Guidance for Biological Criteria (EPA 1990), the requirement for all states to have descriptive biological criteria by 1993 and numerical biological criteria by 1995 (EPA 1990), Ohio's recent adoption of numerical criteria to improve the protection of biological resources (Ohio EPA 1990), the ecological integrity objective in the National Parks Act in Canada, and the adoption of an ecosystem objective in binational programs to protect the Laurentian Great Lakes (Edwards and Regier 1990). Biological criteria (biocriteria) are code words for "indicator of ecological condition" or "biological integrity compared to some least disturbed reference community."

EVALUATING ECOLOGICAL STATUS

Environmental degradation inevitably is tied to a failure of the system to support biological processes at a desired level. Thus, direct assessment of that biology is essential to an evaluation program. Because the current status of these systems is the result of a complex vector of influences, efforts to assess their condition should reflect a vector of attributes. That is, multimetric approaches to biological assessment are both necessary and appropriate to the protection of complex biological systems. When used individually, each metric constitutes a separate test of the system's status and integrative efforts can use *all* the villages of ecology. The final quantitative analysis is more a crude integration of these separate tests than a critical value of inherent merit. Indeed, the independent tests should not be lost in the aggregation because they provide important information on the status of specific attributes of the system.

Advocates of biotic assessment must take special care to avoid the traps associated with dependence on narrow conceptual approaches (such as species richness or thermodynamics) or untested or discredited ecological theories (such as stability/diversity dogma). While indices based on an intuitively satisfying theoretical construct are appealing, their adoption is premature pending extensive testing and validation in real-world situations.

HOW DO WE MEASURE ECOLOGICAL INTEGRITY?

About ten years ago, I was asked by a regional water resource agency to evaluate the conditions (the ecological health or integrity) in the streams in their geographic area of responsibility. As a result, I developed the Index of Biotic Integrity (IBI) as an approach, now widely used in North America and in Europe, to the assessment of the quality of water resources (Karr 1981). Most of the components of the original IBI were well-known associations between system condition and human influence. Using the concept of a biological vector, IBI is an integration of twelve attributes of a segment of the biota (fishes) of a stream. But the sum of these metrics is not an amorphous amalgam as some have suggested, nor is it an ad hoc estimator. Rather, it represents a synthesis of a dozen distinct hypotheses about the relationship between attributes of biological systems under varying influence from human society.

After a careful evaluation of these hypotheses using real data, they were converted to the status of assumptions. It is widely recognized, for example, that species richness (the number of resident native species) declines as human influence increases, as does the number of intolerant and specialist species. In contrast, the number of generalists increases as human influence increases. These patterns have been demonstrated repeatedly for fish and insect assemblages of stream systems (Karr et al. 1986; Ohio EPA 1988; Miller et al. 1988; Lyons 1992; Kerans and Karr 1992) throughout North America. Fragmentation of terrestrial environments also results in declining numbers of resident species of most taxa (Williamson 1981).

These empirical observations, combined with the theoretical underpinnings that generated the original hypotheses, represent a synthesis of ecological information, both theoretical and empirical. They have a basis in our understanding of ecological pattern and process and they have been tested extensively. They are applicable for a diversity of geographic areas and habitat types and for a variety of taxa. Their response range has been verified and documented, and they reflect the reality of natural variation in the attributes of biological systems. That is, by considering the ecological foundation of an IBI metric developed for use in the midwestern United States, an appropriate, ecologically equivalent metric can be developed for use in other regions (Miller et al. 1988).

These observations also demonstrate that biological approaches can provide for informed decisions about the use and protection of natural resources. Ohio, for example, is the first state to incorporate an ecologically robust, biological approach to water resource protection in its water quality regulations. Ohio EPA routinely uses biological monitoring in the application of biological criteria to such diverse activities as use designations, Section 319 Clean Water Act (CWA) non-point-source program activities, Section 305(B) CWA water quality inventory reports, and NPDES discharge permits (Dudley 1991). Ohio's approach, based largely on IBI, is routinely used to guide water policy decisions, to negotiate permit conditions, and even to defend its decisions in judicial proceedings. Finally, at least forty states are working to incorporate biological assessments (IBI or its conceptual clones) in their programs to protect and manage water resources.

CONCLUSION

Human actions threaten the integrity of the biosphere and, there-fore, our welfare. No longer are our impacts short in duration or local in effect; nor can they be treated by conventional medical or technological approaches. The major social and economic transi-tions of the agricultural and industrial revolutions created massive environmental deficits, demonstrating that their blessings have been offset by potentially disastrous consequences.

Our ability to change the world outpaces the ability of biologi-cal systems to respond to these changes. Even our minds and our language evolve too slowly to deal with the rate of environmental degradation. Much of this volume has been devoted to grappling with applying old terminology and old dogma to new challenges. Our success as a world society depends on an environmental revolution—on our ability, that is, to change our vocabulary and our perceptions so that we can indeed protect ecological integrity.

No clear standard has been developed to measure biological condition, however, or to define the nature and extent of degra-dation. The lack of a standard has, by default, resulted in society's ignoring the status of earth's life support systems. No single inno-vation can reverse that trend. Recognizing the problem as a seri-ous social issue fosters dialogue to define our goals—to define the level of ecological health we can accept as a society. Hence we must develop a rational approach to the definition of ecological health, methods to measure that health, and mechanisms to incor-porate protection of ecological health into society's decision-mak-ing processes.

But the diversity of approaches to these issues has resulted in a cacophony of voices that may be more an impediment than a solu-tion to the problem of defining and measuring ecological health. Particularly during this early stage, it is important that we cast the net widely to ensure that no voices and thus no critical approaches are lost. The proximate mechanisms regulating the abundance of species and the attributes of their assemblages result from many weak forces acting probabilistically, the cumulative effects of which are large but the individual effects are minor, interactive, and uncertain (Karr and Freemark 1985). Similarly, the influences of human actions are complex and the responses of the biota vary from system to system.

To protect ecological integrity and thus the integrity of human society, we must first understand the needs of the earth's biota and the needs of human society—needs that are not always convergent but are at least interdependent. Ecologists have been slow to rise to this challenge. And when they have, their conclusions are often colored by their ecological village. If we are to overcome this problem, we must strive to improve our collective ecological literacy. Society's understanding of the consequences of biotic impoverishment, of the extent to which we depend on healthy biological systems, is rudimentary. A primary focus on biotic impoverishment as the problem and the protection of ecological or biological integrity as the solution will draw attention to the critical importance of biological sustainability.

References

Botkin, D. 1990. *Discordant Harmonies: A New Ecology for the Twenty-first Century*. New York: Oxford University Press.

Callicott, J. B. 1991. "Conservation Ethics and Fishery Management." *Fisheries* (Bethesda) 16(2): 22–28.

Costle, D. M. 1979. *Saving Ourselves Broke: The Future Expenses of Deferred Regulation*. OPA 75/9. Washington: EPA.

Dudley, D. R. 1991. "A State Perspective on Biological Criteria in Regulation." In *Biological Criteria: Research and Regulation*. EPA-440/5-91-005. Washington: EPA.

Edwards, C. J., and H. A. Regier (eds.). 1990. *An Ecosystem Approach to the Integrity of the Great Lakes in Turbulent Times*. Special Publication 90-94. Ann Arbor, Mich.: Great Lakes Fishery Commission.

Holling, C. S. 1986. "The Resilience of Terrestrial Ecosystems: Local Surprise and Global Change." In W. C. Clark and R. E. Munn (eds.), *Sustainable Development of the Biosphere*. Cambridge: Cambridge University Press.

Karr, J. R. 1981. "Assessment of Biotic Integrity Using Fish Communities." *Fisheries* (Bethesda) 6(6):21–27.

———. 1991. "Biological Integrity: A Long-Neglected Aspect of Water Resource Management." *Ecological Applications* 1:66–84.

———. 1992. "Lessons from Practical Attempts to Measure Biotic Integrity." In S. Woodley, G. Francis, and J. Kay (eds.), *Ecological Integrity and the Management of Ecosystems*. Ann Arbor, Mich.: Lewis.

Karr, J. R., and D. R. Dudley. 1981. "Ecological Perspective on Water Quality Goals." *Environmental Management* 5:55–68.

Karr, J. R., and K. E. Freemark. 1985. "Disturbance and Vertebrates: An Integrative Perspective." In S.T.A. Pickett and P. S. White (eds.), *The Ecology of Natural Disturbance and Patch Dynamics*. New York: Academic Press.

Karr, J. R., L. A. Toth, and D. R. Dudley. 1985. "Fish Communities of Midwestern Rivers: A History of Degradation." *BioScience* 35:90–95.

Karr, J. R. et al. 1986. *Assessment of Biological Integrity in Running Water: A Method and Its Rationale*. Publication 5. Champaign: Illinois Natural History Survey.

Kerans, B. L., and J. R. Karr. 1992. "Assessment of Biological Integrity Using Communities of Benthic Invertebrates." Unpublished manuscript.

Leopold, A. 1949. *A Sand County Almanac and Sketches Here and There.* New York: Oxford University Press.

Lyons, J. 1992. *Using the Index of Biotic Integrity (IBI) to Measure Environmental Quality in Warmwater Streams of Wisconsin*. General Technical Report NC-149. St. Paul, Minnesota: North Central Forest Experiment Station, U.S. Forest Service.

Meadows, D. H. 1990. "A Reaction from a Multitude." In G. M. Woodwell (ed.), *The Earth in Transition: Patterns and Processes of Biotic Impoverishment*. Cambridge: Cambridge University Press.

Miller, D. L. et al. 1988. "Regional Applications of an Index of Biotic Integrity for Use in Water Resource Management." *Fisheries* (Bethesda) 13(5):12–20.

Nehlsen, W., J. E. Williams, and J. A. Lichatowich. 1991. "Pacific Salmon at the Crossroads: Stocks at Risk from California, Oregon, Idaho, and Washington." *Fisheries* (Bethesda) 16(2):4–21.

Norton, B. G. 1991. "Ecological Health and Sustainable Resource Management." In R. Costanza (ed.), *Ecological Economics: The Science and Management of Sustainability*. New York: Columbia University Press.

Ohio EPA. 1988. *Users Manual for Biological Field Assessment of Ohio Surface Waters*. 3 vols. Columbus: Ohio Environmental Protection Agency.

———. 1990. *The Use of Biocriteria in the Ohio EPA Surface Water Monitoring and Assessment Program*. Columbus: Ohio Environmental Protection Agency, Division of Water Planning and Assessment.

Ornstein, R., and P. Ehrlich. 1989. *New World, New Mind: Moving Toward Conscious Evolution*. New York: Doubleday.

Pimm, S. L. 1984. "The Complexity and Stability of Ecosystems." *Nature* 307:321–326.

Schoener, T. 1986. "Overview: Kinds of Ecological Communities—Ecology Becomes Pluralistic." In J. M. Diamond and T. J. Case (eds.), *Community Ecology*. New York: Harper & Row.

Science Advisory Board (SAB). 1990. *Reducing Risk: Setting Priorities and Strategies for Environmental Protection*. SAB-EC-90-021. Washington: EPA.

Studds, G. E. 1991. "Preserving Biodiversity." *BioScience* 41:602.

Suter, G. W., II. 1991. "Is Ecosystem Health More Than a Metaphor?" Abstracts of the 12th Annual Meeting. Seattle: Society for Environmental Toxicology and Chemistry.

U. S. Environmental Protection Agency (EPA). 1990. *Biological Criteria: National Program Guidance for Surface Waters*. EPA 440/5-90-004. Washington: EPA, Office of Water.

Williams, J. E. et al. 1989. "Fishes of North America Endangered, Threatened, or of Special Concern: 1989." *Fisheries* (Bethesda) 14(6):2–20.

Williamson, M. 1981. *Island Populations*. Oxford: Oxford University Press.

Woodwell, G. M. 1989. "On Causes of Biotic Impoverishment." *Ecology* 70:14–15.

———. (ed.). 1990. *The Earth in Transition: Patterns and Processes of Biotic Impoverishment*. Cambridge: Cambridge University Press.

14

Toward an Operational Definition of Ecosystem Health

ROBERT COSTANZA

Ecosystem health is a normative concept: a bottom line. It represents a desired endpoint of environmental management, but the concept has been difficult to use because of the complex, hierarchical nature of ecosystems. Without an adequate operational definition of the desired endpoint, effective management is unlikely. This chapter investigates some of the operational definitions of ecosystem health, weighs their advantages and disadvantages, suggests alternatives that synthesize these past definitions, and suggests research paths aimed at allowing these new definitions to be put into practice.

Concept definitions of ecosystem health can be summarized as:

- Health as homeostasis
- Health as the absence of disease
- Health as diversity or complexity
- Health as stability or resilience
- Health as vigor or scope for growth
- Health as balance between system components

All of these concepts represent pieces of the puzzle, but none is comprehensive enough to serve our purposes here. Hence I wish to elaborate on the concept of ecosystem health as a comprehensive, multiscale, dynamic, hierarchical measure of system resilience, organization, and vigor. These concepts are embodied in the term "sustainability," which implies the system's ability to maintain its structure (organization) and function (vigor) over time in the face of external stress (resilience). A healthy system must also be defined in light of both its context (the larger system of which it is part) and its components (the smaller systems that make it up).

WHY DO WE NEED CONCEPTS FOR MANAGING ECOSYSTEMS?

All complex systems are, by definition, made up of a number of interacting parts. In general, these components vary in their type, structure, and function within the whole system. Thus a system's behavior cannot be summarized simply by adding up the behavior of the individual parts. Contrast a simple physical system (say an ideal gas) with a complex biological system (say an organism). The temperature of the gas is a simple aggregation of the kinetic energy of all the individual molecules in the gas. The temperature, pressure, and volume of the gas are related by simple relationships with little or no uncertainty. An organism, however, is composed of complex cells and organ systems. The state of an organism cannot be surmised simply by adding up the states of the individual components, since these components are themselves complex and have different, noncommensurable functions within the overall system. Indicators that might be useful for understanding heart function—pumping rate and blood pressure, for instance—are meaningless for skin or teeth.

But to understand and manage complex systems, we need some way of assessing the system's overall performance (its relative "health"). The EPA has recently begun to shift the stated goals of its monitoring and enforcement activities from protecting only "human health" to protecting overall "ecological health." Indeed, EPA's Science Advisory Board (SAB 1990:17) recently stated:

EPA should attach as much importance to reducing ecological risk as it does to reducing human health risk. These very close linkages between human health and ecological health should be reflected in national environmental policy. When EPA compares the risks posed by different environmental problems in order to set priorities for Agency action, the risks posed to ecological systems must be an important part of the equation.

Although this statement gives the concept of ecological health importance as a primary EPA goal, it begs the question of what ecosystem health *is*, while tacitly defining it as analogous to human health. The dictionary definitions of health are: "1. the condition of being sound in mind, body, and spirit; 2. flourishing condition or well-being." These definitions are rather vague. In order to meet the mandate for effectively managing the environment we must construct a more rigorous and operational definition of health that is applicable to all complex systems at all levels of scale, including organisms, ecosystems, and economic systems. As the preceding chapters make clear, there is a wide range of opinion about how to proceed with this daunting task. Some interesting possibilites have been proposed, however, and I think we are off to a good start.

In its simplest terms, then, health is a measure of the overall performance of a complex system that is built up from the behavior of its parts. Such measures of system health imply a *weighted* summation or a more complex operation over the component parts, where the weighting factors incorporate an assessment of the relative importance of each component to the functioning of the whole. This assessment of relative importance incorporates "values," which can range from subjective and qualitative to objective and quantitative as we gain more knowledge about the system under study. In the practice of human medicine, these weighting factors or values are contained in the body of knowledge and experience embodied in the medical practitioner.

Figure 1 shows the progression from directly measured "indicators" of a component's status, through "endpoints" that are composites of these indicators, to health with the help of "values." Measures of health are inherently more difficult, more comprehensive, require more modeling and synthesis, and involve less precision, but are more relevant than the endpoints and indicators from which they are built. It remains to determine which general approaches to developing these measures of health for ecosystems are most effective. I begin with a review of current approaches and end with a synthesis incorporating the best features of these systems along with some new ideas.

CURRENT CONCEPTS OF ECOSYSTEM HEALTH

Concepts of ecosystem health can be derived by analogy with concepts of human health (Rapport 1989; Chapter 8 of this volume). Both human individuals and ecosystems are complex systems composed of interacting parts in a complex balance of interdependent function. But humans are warm-blooded, homeostatic systems and human physicians have a compendium of known

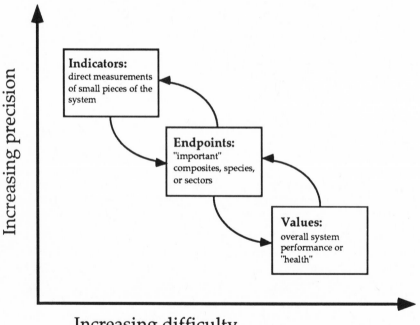

FIGURE 1. *Relationship of indicators, endpoints, and values*

diseases, a wide body of reference data on the "standard human," and many diagnostic tools (Schaeffer et al. 1988; Chapter 9 of this volume). None of these aids are available for ecosystems. There is, however, a large body of literature on the related concepts of ecosystem stability and resilience (reviewed in Pimm 1984 and Holling 1986). In the following discussion I make use of several terms and concepts that have emerged in this ecological literature over the past forty years. They are summarized and defined in Table 1 (an expanded version of a table appearing in Pimm 1984).

What we are after is a general concept of *complex system health* that draws on ideas from human health practice and ecosystem (and economic system) theory and practice but is equally applicable to evaluating the health of any complex system at any scale—from cells to organs to organisms to populations to ecosystems and economic systems.

Health as Homeostasis

The simplest and most popular definition of system health is health as homeostasis: Any and all changes in the system represent a decrease in health. It is popular because it does not require a weighted synthesis of the raw indicators. If any one indicator is seen to change beyond the range of "normal variation," the system's health is deemed to have suffered. The only complication is differentiating internal natural variation from that induced by outside stress. This approach works moderately well for organisms, especially warm-blooded vertebrates, since they are homeostatic and there is a very large population from which to determine "normal ranges." But for ecosystems, economic systems, and other nonhomeostatic systems with small populations, it works less well or not at all.

First of all, we cannot assume that all change is bad (or at least not equally bad). Even in homeostatic organisms one must be able to attach some relative importance (based on overall system function or health) to each indicator in order to get any idea of *how*

TABLE 1. *Definitions of Key Variables*

Variable	Definition
Stability	
Homeostasis	Maintenance of steady state in living organisms by use of feedback control processes.
Stable	A system is stable if and only if variables all return to initial equilibrium after being disturbed. A system is locally stable if this return applies to small perturbations and globally stable if it applies to all possible perturbations.
Sustainable	A system's ability to maintain structure and function indefinitely. All nonsuccessional (climax) ecosystems are sustainable, but they may not be stable. (See Resilience below.) Sustainability is a policy goal for economic systems.
Resilience	1. How fast the variables return to equilibrium following perturbation; not defined for unstable systems (Pimm 1984). 2. A system's ability to maintain structure and patterns of behavior in the face of disturbance (Holling 1986).
Resistance	The degree to which a variable is changed following perturbation.
Variability	The variance of population densities over time or allied measures such as standard deviation or coefficient of variation (SD/mean).
Complexity	
Species richness	The number of species in a system.
Connectance	The number of interspecific interactions divided by possible interspecific interactions.
Interaction strength	The mean magnitude of interspecific interaction: size of the effect of one species' density on growth rate of another species.
Evenness	The variance of species abundance distribution.
Diversity indices	Measures combining evenness and richness with a particular weighting for each; one important member of this family is the information-theory index, H.
Ascendency	An information-theory measure combining average mutual information (a measure of connectedness) and the system's total throughput as a scaling factor (see Chapter 11).
Others	
Perturbation	Change to a system's inputs or environment beyond normal range of variation.
Stress	Perturbation with negative effect on a system.
Subsidy	Perturbation with a positive effect on a system.

Source: Adapted from Pimm (1984) and Holling (1986).

much health has changed. Gaining ten pounds is not as bad as developing a heart murmur which is not as bad as developing cancer, even though all of these can be considered departures from homeostasis. Moreover ecosystems are complicated by the process of succession. If conditions change sufficiently, there will be replacement of one ecosystem with another that is better adapted to the new conditions. This represents radical change for the first system, and unless one can assess the relative value of the larger system before and after succession, then all successional changes must be considered harmful. But nature (even with no human intervention) is in a constant state of adjustment, change, and succession, not in a state of stasis or equilibrium. We must therefore look a little deeper for an adequate definition of ecosystem health.

Health as Absence of Disease

Health is often defined as the absence of disease. But to operationalize this definition one must first define disease. The dictionary defines it as "any departure from health." Thus defining health as absence of disease just restates the problem in the negative. If we can define health, we can also define nonhealth or disease and vice versa. It also casts the problem in terms that are much too black or white. We desire a continuous measure of relative health, not a binary index of healthy or not healthy.

For organisms, a more useful definition of disease is possible: "a particular destructive process in the body, with a specific cause and characteristic symptoms; a specific illness, ailment, or malady." Disease can thus be thought of as a stress to the system, a perturbation with certain negative effects. One can catalog and eliminate all anthropogenic stresses on an ecosystem, but without an independent definition of health it is impossible to know which of these stresses really cause problems (are negative versus positive) and to what degree. If one uses the homeostatic definition of health, then any stress that causes any change in the system is a disease—and this seems much too severe and unrealistic. In fact, as C. S. Holling (1986, 1992) points out, ecosystem health may be related more to the ability of systems to *use* stress creatively than to their ability to resist it completely. More on this later.

Health as Diversity or Complexity

A third possible definition of ecosystem health is linked to the system's diversity or complexity. The idea is that diversity or complexity are predictors of stability or resilience and that these are measures of health. (See Table 1.) This linkage has been a subject of much controversy in the ecological literature, and the tide has turned several times. But because diversity is so easy to measure in ecosystems it has come to be a prime *de facto* indicator of health. According to Stewart Pimm (1984) there are several interesting aspects of the problem that have yet to be investigated. Recent advances in network analysis (Wulff et al. 1989; Chapter 11 of this volume) hold some promise in allowing a more sophisticated view of the *organization* of systems, not just their number of parts as in diversity measures. More on this later.

Health as Stability or Resilience

Stability and the related concept of resilience have much to recommend them as general measures of health. Healthy organisms have the ability to withstand disease organisms. They are resilient and recover quickly after a perturbation. Hence this leads to a definition of health as the ability to recover from stress. The greater this ability the healthier the system. A problem with this definition is that it says nothing about the system's operating level or degree of organization. A dead system is more stable than a live system because it is more resistant to change. But it is certainly not healthier. Thus an adequate definition of health should also say something about the system's level of activity and organization as well. There is a related, but in many respects more appealing definition of resilience as "the ability of a system to maintain its structure and patterns of behavior in the face of disturbance" (Holling 1986). This definition stresses the adaptive nature of ecosystems rather than the speed with which they can shrug off perturbations. Systems are healthy if they can absorb stress and use it creatively rather than simply resisting it and maintaining their former configurations. Southern pine forests, for example, are adapted to deal with frequent fires as a necessary part of their overall functioning. Efforts to suppress this stress are counterproductive and lead to larger, more destructive fires in the long run.

Health as Vigor or Scope for Growth

It has been hypothesized that a system's ability to recover from stress (or to utilize it) is related to its overall metabolism or energy flow (Odum 1971) or to its "scope for growth" (Bayne et al. 1987)—the difference between the energy required for system maintenance and the energy available to the system for all purposes. Both measures attempt to get at the system's capability to respond to generalized stress as well as its overall level of activity and organization and thus are one step deeper than stability or resilience alone. They are intended to be predictors of activity-weighted resilience and, ultimately, health.

Health as Balance between Components

Another concept that has wide acceptance in Eastern traditional medicine is the idea that a healthy system is one that maintains the proper balance between system components. This idea of balance is deeply ingrained in ecological theory as well, but it has usually been used as a general explanation for existing distributions (the ecosystem is in balance) rather in any predictive or diagnostic way. How do we know if the system is out of balance unless we have some overall indicator of health against which to judge?

TOWARD A PRACTICAL DEFINITION

How can we create a practical definition of system health that is applicable with equal facility to complex systems at all scales? Let us first lay out the minimum characteristics of such a definition. First, an adequate definition of ecosystem health should integrate the concepts of health mentioned above. Specifically it should be a combined measure of system resilience, balance, organization (diversity), and vigor (metabolism). Second, the definition should be a comprehensive description of the system. Looking at only one part of the system implicitly gives the remaining parts zero weight. Third, the definition will require the use of weighting factors to compare and aggregate different components in the system. It should use weights for components that are linked to the functional dependence of the system's sustainability on the components, and the weights should be able to vary as the system changes to account for "balance." And fourth, the definition should be hierarchical to account for the interdependence of various time and space scales.

At a recent workshop attended by a wide range of participants including scientists, philosophers, managers, environmentalists, and industry representatives the following working definition of ecosystem health was arrived at (see the Introduction to this volume): An ecological system is healthy and free from "distress syndrome" if it is stable and sustainable—that is, if it is active and maintains its organization and autonomy over time and is resilient to stress. Ecosystem health is thus closely linked to the idea of sustainability, which is seen to be a comprehensive, multiscale, dynamic measure of system resilience, organization, and vigor. This definition is applicable to all complex systems from cells to ecosystems to economic systems (hence it is comprehensive and multiscale) and allows for the fact that systems may be growing and developing as a result of both natural and cultural influences. According to this definition, a diseased system is one that is not sustainable and will eventually cease to exist. The time and space frame are obviously important in this definition. Individual organisms are not sustainable indefinitely, but the populations and ecosystems of which they are part may be sustainable indefinitely. Distress syndrome refers to the irreversible processes of system breakdown leading to death (Rapport 1981). To be healthy and sustainable, a system must maintain its metabolic activity level as well as its internal structure and organization (a diversity of processes effectively linked to one another) and must be resilient to outside stresses over a time and space frame relevant to that system.

What does this mean in practice? Table 2 lays out the three main components of this proposed concept of system health (resilience, organization, and vigor) along with related concepts and measurements in various fields.

The following preliminary form for an overall system health index (HI) is proposed:

$$HI = V^*O^*R$$

where: HI = system health index, also a measure of sustainability

TABLE 2. *Indices of Vigor, Organization, and Resilience in Various Fields*

Component of health	Related concepts	Related measures	Field of origin	Probable method of solution
Vigor	Function Productivity Throughput	GPP, NPP, GEP GNP Metabolism	Ecology Economics Biology	Measurement
Organization	Structure Biodiversity	Diversity index Average mutual in- formation predictability	Ecology Ecology	Network analysis
Resilience		Scope for growth	Ecology	Simulation modeling
Combinations		Ascendancy	Ecology	

V = system vigor, a cardinal measure of system activity, metabolism, or primary productivity

O = system organization index, a 0–1 index of the relative degree of the system's organization, including its diversity and connectivity

R = system resilience index, a 0–1 index of the relative degree of the system's resilience

This formulation leads to a comprehensive index incorporating the three major components outlined above. In essence, it is the system's vigor or activity weighted by indices for relative organization and resilience. In this context, eutrophication is unhealthy in that it usually represents an increase in metabolism that is more than outweighed by a decrease in organization and resilience. Artificially eutrophic systems tend toward lower species diversity, shorter food chains, and lower resilience.

Naturally eutrophic systems have developed higher diversity and organization along with higher metabolism and are therefore healthier. To operationalize these concepts (especially organization and resilience) the *HI* will require a heavy dose of systems modeling. We turn now to a summary of the available tools.

Network Analysis

Ecology is often defined as the study of the relationships between organisms and their environment. The quantitative analysis of interconnections between species and their abiotic environment has therefore been a central issue. The mathematical analysis of interconnections is important in other fields as well. Practical quantitative analysis of interconnections in complex systems began with the economist Wassily Leontief (1941) using what has come to be called input/output (I-O) analysis. Recently these concepts, sometimes called flow analysis, have been applied to the study of interconnections in ecosystems (Hannon 1973, 1976, 1979, 1985a, 1985b, 1985c; Costanza and Neill 1984). Related ideas were developed from a different perspective in ecology, under the heading of compartmental analysis (Barber et al. 1979; Finn 1976; Funderlic and Heath 1971). Walter Isard (1972) was the first to attempt a combined ecological/economic system I-O analysis, and combined ecological/economic mass-balance models have been proposed by several others (Daly 1968; Cumberland 1987). We refer to the total of all variations of the analysis of ecological and economic networks as *network analysis*.

Network analysis holds the promise of allowing an integrated, quantitative, hierarchical treatment of all complex systems, including ecosystems and combined ecological/economic systems. One promising route is the use of "ascendancy" (Ulanowicz 1980, 1986) and related measures (Wulff et al. 1989; Chapter 11 of this volume) to measure the degree of organization in ecological, economic, or any other networks. Measures like ascendency go several steps beyond the traditional diversity indices used in ecology. They estimate not only how many different species there are in a system but, more important, how those species are organized. This kind of measure may provide a necessary input for a quantitative and general index of system health applicable to both ecological and economic systems.

Another promising avenue of research in network analysis has to do with its use for "pricing" commodities in ecological or economic systems. The mixed-units problem arises in any field that tries to analyze interdependence in complex systems that have many different types and qualities of interacting parts. Ecology and economics are two such fields. Network analysis in ecology has avoided this problem in the past by arbitrarily choosing one commodity flowing through the system as an index of interdependence (carbon, enthalpy, nitrogen, and the like). This strategy ignores the interdependencies between commodities and assumes that the chosen commodity is a valid tracer for relative value or importance in the system. This assumption is unrealistic and severely limits the comprehensiveness of an analysis whose major objective is to deal comprehensively with whole systems.

There are evolving methods for dealing with the mixed-units problem based on analogies to the calculation of prices in economic input/output models. Starting with a more realistic commodity-by-process description of ecosystem networks that allows for joint products, one can use *energy intensities* to ultimately convert the multiple commodity description into a pair of matrices that can serve as the input for standard (single-commodity) network analysis. The new single-commodity description incorporates commodity and process interdependencies in a manner analogous to the way economic value incorporates production interdependencies in economic systems (Costanza and Hannon 1989; Chapter 12 of this volume). This analysis would allow objective valuation of components of ecosystems and combined ecological/economic systems as a complement to subjective evaluations and better overall measures of *vigor* and *organization* in the system.

Simulation Modeling

Evaluating the health of complex systems demands a pluralistic approach (Norgaard 1989, Rapport 1989) and an ability to integrate and synthesize the many different perspectives that can be taken. Probably no single approach or paradigm is sufficient because, like the blind men and the elephant, the subject is too big and complex to touch it all with one limited set of perceptual tools. Rather, we must extend our view to cover the diversity of approaches that may shed light on the problem and must also develop the ability to use *all* of the available light to understand the system.

We need an integrated, multiscale, transdisciplinary, and pluralistic approach to quantitative modeling of systems (including organisms, ecosystems, and ecological/economic systems). While this approach has frequently been suggested (Norgaard 1989), it is difficult to operationalize with traditional funding mechanisms. Such an approach would allow the relationships between scales and modeling approaches to be directly investigated and would result in a deeper understanding of the systems under study. It would produce new ways of *scaling*—or using information at one scale to build understanding and models at other scales. This kind of scaling is essential to developing quantitative measures of system health, especially in addressing the resilience component of the system health index proposed here.

Quantifying resilience implies an ability to predict the dynamics of the system under stress. Predicting these ecosystem impacts requires sophisticated computer simulation models that represent a synthesis of the best available understanding of the way these complex systems function (Costanza et al. 1990). Beyond its use in health indices, this modeling capability is essential for regional ecosystem management and also for modeling the response of regional and global ecosystems to regional and global climate change, sea level rise, acid precipitation, toxic waste dumping, and a host of other potential impacts. Several recent developments make this kind of modeling feasible—not only the improved accessibility of extensive spatial and temporal data bases from remote sensing, aerial photography, and other sources (including EPA's new EMAP program), but also advances in computer power and convenience that make it possible to build and run predictive models at the necessary levels of spatial and temporal resolution.

CONCLUSION

There is no silver bullet that will allow us to assess ecosystem health quickly, cheaply, precisely, and without ambiguity. There is no health meter with probes that can be inserted into ecosystems to yield a digital readout of health. Assessing health in a complex system—from organisms to ecosystems to economic systems—requires a good measure of judgment, precaution, and humility, but also a good measure of systems analysis and modeling in order to put all the individual pieces together into a coherent picture. Human health assessment likewise requires systems analysis, but the compendium of known diseases available, the huge body of reference data on the "standard human," and many types of diagnostic tools available make human health assessment possible without resort to sophisticated computer modeling. Not that it does not require sophisticated analysis, but this analysis is embodied in the "expert system" of medical practitioners. To generalize this expertise and apply it to all kinds of systems requires systems modeling.

I have proposed a general index of system health made up of three components: vigor, organization, and resilience. Vigor can be measured directly in most cases. I have suggested that network analysis and simulation modeling are two of the most promising avenues for the development of the organization and resilience components of the proposed index. Both are relatively expensive to implement because of their large data requirements. But developments like EPA's Environmental Monitoring and Assessment Program (EMAP) that are designed to collect much of the data necessary to implement these analyses, along with remote sensing data and more capable and friendly computers, make implementing these methods finally feasible.

The idea is also amenable to direct empirical testing. One approach might be to apply various versions of the health index to systems for which we already have general agreement on health status—for example, humans or other organisms. One could then see which version did the best job of replicating our agreed-on health rankings in these systems as a test of their effectiveness. The best indices could then be applied to ecological systems with at least some confidence that they do represent the general health of the system. None of this will be easy or simple, but it is time to begin the messy, difficult, and absolutely essential task of assessing the health of ecological systems.

References

Barber, M., B. Patten, and J. Finn. 1979. "Review and Evaluation of I-O Flow Analysis for Ecological Applications." In J. Matis, B. Patten and G. 'White (eds.), *Compartmental Analysis of Ecosystem Models*. Vol. 10 of *Statistical Ecology*. Bertonsville, Md.: International Cooperative Publishing House.

Bayne, B. L. et al. 1987. *The Effects of Stress and Pollution on Marine Animals*. New York: Praeger.

Costanza, R. 1987a. "Simulation Modeling on the Macintosh Using STELLA." *BioScience* 37:129–132.

254   SCIENCE AND POLICY

———. 1987b. "Social Traps and Environmental Policy." *BioScience.* 37:407–412.

———. 1989a. "What is Ecological Economics?" *Ecological Economics* 1:1–7.

———. 1989B. "Model Goodness of Fit: A Multiple Resolution Procedure." *Ecological Modelling* 47:199–215.

Costanza, R., and H. E. Daly. 1987. "Toward an Ecological Economics." *Ecological Modelling* 38:1–7.

Costanza, R., and B. M. Hannon. 1989 "Dealing with the 'Mixed Units' Problem in Ecosystem Network Analysis." In F. Wulff, J. G. Field, and K. H. Mann (eds.), *Network Analysis of Marine Ecosystems: Methods and Applications.* Coastal and Estuarine Studies Series. Heidelberg: Springer-Verlag.

Costanza, R., and C. Neill. 1984. "Energy Intensities, Interdependence, and Value in Ecological Systems: A Linear Programming Approach." *Journal of Theoretical Biology* 106:41–57.

Costanza, R., and F. H. Sklar. 1985. "Articulation, Accuracy, and Effectiveness of Mathematical Models: A Review of Freshwater Wetland Applications." *Ecological Modelling* 27:45–68

Costanza, R., S. C. Farber, and J. Maxwell. 1989. "The Valuation and Management of Wetland Ecosystems." *Ecological Economics* 1:335–361.

Costanza, R., F. H. Sklar, and M. L. White. 1990. "Modeling Coastal Landscape Dynamics." *BioScience* 40:91–107.

Costanza, R., M. H. Sagoff, J. A. Bartholomew, and B. Haskell. 1991. *Summary Report: Workshop on Ecosystem Health and Environmental Management.* College Park, Md.: Maryland Sea Grant.

Cumberland, J. H. 1987. "Need Economic Development Be Hazardous to the Health of the Chesapeake Bay?" *Marine Resource Economics.* 4:81–93.

Daly, H. 1968. "On Economics as a Life Science." *Journal of Political Economy.* 76:392–406.

Farber, S., and R. Costanza. 1987. "The Economic Value of Wetlands Systems." *Journal of Environmental Management* 24:41–51.

Finn, J. 1976. "The Cycling Index." *Journal of Theoretical Biology* 56:363–73.

Funderlic, R., and M. Heath. 1971. *Linear Compartmental Analysis of Ecosystems.* ORNL-IBP-71-4. Oak Ridge, Tenn.: Oak Ridge National Laboratory.

Hannon, B. 1973. "The Structure of Ecosystems." *Journal of Theoretical Biology* 41:535–46.

———. 1976. "Marginal Product Pricing in the Ecosystem." *Journal of Theoretical Biology* 56:256–267.

———. 1979. "Total Energy Costs in Ecosystems." *Journal of Theoretical Biology* 80:271–293.

―――. 1985a. "Ecosystem Flow Analysis." *Canadian Bulletin of Fisheries and Aquatic Sciences* 213:97–118.

―――. 1985b. "Conditioning the Ecosystem." *Mathematical Biology* 75:23–42.

―――. 1985c. "Linear Dynamic Ecosystems." *Journal of Theoretical Biology* 116:89–98.

Hannon, B., R. Costanza, and R. A. Herendeen. 1986. "Measures of Energy Cost and Value in Ecosystems." *Journal of Environmental Economics and Management* 13:391–401.

Hannon, B., R. Costanza, and R. Ulanowicz. 1991. "A General Accounting Framework for Ecological Systems." *Theoretical Population Biology* 40:78-104.

Holling, C. S. 1973. "Resilience and Stability of Ecological Systems." *Annual Review of Ecology and Systematics* 4:1–23.

―――. 1986. "The Resilience of Terrestrial Ecosystems: Local Surprise and Global Change." In W. C. Clark and R. E. Munn (eds.), *Sustainable Development of the Biosphere*. Cambridge: Cambridge University Press.

―――. 1992. "The Role of Forest Insects in Structuring the Boreal Forest." In H. H. Shugart (ed.), *A Systems Analysis of the Global Boreal Forest*. Cambridge: Cambridge University Press.

Isard, W. 1972. *Ecologic-Economic Analysis for Regional Development*. New York: Free Press.

Karr, J. R. 1981. "Assessment of Biotic Integrity Using Fish Communities." *Fisheries* 6:21–27.

Leontief, W. 1941. *The Structure of American Economy, 1919–1939*. New York: Oxford University Press.

May, R. M. 1973. *Stability and Complexity in Model Ecosystems*. Princeton: Princeton University Press.

―――. 1977. "Thresholds and Breakpoints in Ecosystems with a Multiplicity of Stable States." *Nature* 269:471–477.

Norgaard, R. B. 1989. "The case for Methodological Pluralism." *Ecological Economics* 1:37–57.

Odum, H. T. 1971. *Environment, Power and Society*. New York: Wiley.

O'Neill, R. V., D. L. DeAngelis, J. B. Waide, and T.F.H. Allen. 1986. *A Hierarchial Concept of Ecosystems*. Princeton: Princeton University Press.

Pimm, S. L. 1984. "The Complexity and Stability of Ecosystems." *Nature* 307:321–326.

Rapport, D. J. 1989. "What Constitutes Ecosystem Health?" *Perspectives in Biology and Medicine* 33:120–132.

Rapport, D. J., H. A. Regier, and T. C. Hutchinson. 1985. "Ecosystem Behavior Under Stress." *American Naturalist* 125:617–640.

Schaeffer, D. J., E. E. Herricks, and H. W. Kerster. 1988. "Ecosystem Health: 1. Measuring Ecosystem Health." *Environmental Management* 12:445–455.

Science Advisory Board (SAB). 1990. *Reducing Risk: Setting Priorities and Strategies for Environmental Protection.* SAB-EC-90-021. Washington: EPA.

Turner, M. G., R. Costanza, and F. H. Sklar. 1989. "Methods to Compare Spatial Patterns for Landscape Modeling and Analysis." *Ecological Modelling* 48:1–18.

Turner, M. G., R. Costanza, T. M. Springer, and E. P. Odum. 1988. "Market and Nonmarket Values of the Georgia Landscape." *Environmental Management* 12:209–217.

Ulanowicz, R. E. 1980. "An Hypothesis on the Development of Natural Communities." *Journal of Theoretical Biology* 85: 223–245.

———. 1986. *Growth and Development: Ecosystems Phenomenology.* New York: Springer-Verlag.

Wulff, F., J. G. Field, and K. H. Mann. 1989. *Network Analysis of Marine Ecosystems: Methods and Applications.* Coastal and Estuarine Studies Series. Heidelberg: Springer-Verlag.

Index

Abrams Creek, 186
Absence of disease, 4, 245
Accountability, 106
Accounting framework, 214
Acid precipitation, 252
Acid rain, 5, 163
Aerial photos, 209
Aesthetic value, 58
Agrarian, 219
Aldo Leopold, 47, 100, 124, 174, 225
American Indians, 45
Antarctica, 45
Anthropogenic
    changes, 47
    disturbance, 138
    stress, 140
    stresses, 245
Antiquarians, 46
Antireductionistic, 16
Appalachian,183, 184
Aquatic Ecosystem Health and Management Society, 144
Aristotle, 31, 45, 73, 74, 146
Army Corps of Engineers, 79
Arsenic, 11
Artificial species, 64
Ascendency, 190, 193, 194, 195, 197, 203, 230, 250
Assimilative capacity, 34
Astronomical, 45
Atomism, 23, 27, 29
Atwood's machine, 196
Autocatalysis, 194
Autonomous, 4
Autopoiesis, 51, 52

Average mutual information, 193
Axiom, 6
    of Creativity, 25
    of Differential Fragility, 26, 36, 37
    of Dynamism, 25
    of Hierarchy, 25
    of Relatedness, 25
    of ecological management, 25

Baltic Sea, 12
Bambi syndrome, 127
Barry Commoner, 126
Behavioral psychologists, 62
Bioconservatives, 60
Biocriteria, 232
Biodiversity, 44, 225, 229
    bill, 226
Biological
    conservation, 44
    integrity, 59
    stressors, 164
Biomedical model, 149
Bioregion, 36
Biospheric support systems, 66
Biotechnologists, 60, 63
Biotechnology, 60
Biotic communities, 45
Biotic right, 54
Birth rate, 207
Blue Ridge Parkway, 172, 183
Boreal forest, 11
Boundary issues, 33
Brazil, 68
Business Week, 228

Camus, 114
Cape Hatteras, 172, 186
Capital, 209
Capital replacement rate, 215
Captain John Smith, 67
Carbon, 220
    cycling, 200
Carnivores, 214
Carrying capacity, 66
Catastrophe theory, 139
Central European forests, 11
Charles Darwin, 59
Chemical stressors, 164
Chesapeake Bay, 6, 36, 63, 67,
    70, 197
Child care, 212
China clay waste, 12
Chronic disease, 212
Civilian Conservation Corps,
    171
Clean Air Amendments, 161
Clean Water Act, 226, 232,
    234
Clements, F. E. , 136
Clementsian holistic paradigm,
    46
Clementsian view, 51, 171,
    229
Climate change, 146
Climax community, 171, 191,
    192
Climax species, 218
Climax theory, 171, 172
Clinical ecology, 144
Coal surface mining, 11
Compartmental analysis, 250
Compensation test, 103, 104
Complex system health, 243
Complexity, 26, 50, 59, 112,
    117, 170, 239, 246
Concepts of valuation, 97

Connectance, 230
Conservationist, 47
Constructivism, 35
Consumption rate, 207
Consumptive preferences, 4
Contextualism, 27
Contextualist paradigm, 26
Continuous measure of rela-
tive health, 245
Coral reef, 12
Corporate, 210
Cost-benefit analysis, 101,
    109, 111
Counteractive capacity, 150
Craggy Pinnacles, 183
Creativity, 6
Critical Trends Assessment
Program (CTAP), 167
Crystal River, Florida, 190,
    196
Cultural
    home, 145
    relativism, 99
    -boundedness, 99
Cumberland Island, 178, 185,
Curtis Prairie, 176
Cycling and feedback, 8

Darwin, 44
Darwinian evolution, 27
Decomposers, 214
Deer irruptions, 48
Deflation, 211
Deforestation, 11
Democratic polity, 46
Democritus, 31
Depreciation, 210
Descartes, 31
Detritivory/herbivory ratio,
    201
Diagnosis., 10

Diagnostic
   protocols, 151, 152
   tools, 14
Differential Fragility, 6
Discount, 211
Diseases, 10
Dispersal, 231
Dissipative structures, 15
Distributional consequences,
   104, 106
Disturbance theory, 138
Disturbances, 214
Diversity, 49, 50, 59, 137, 170,
   171, 175, 176, 182, 184, 185,
   186, 187, 192, 195, 203, 216,
   223, 231, 233, 239, 246, 247,
   249, 250
   -stability hypothesis, 44,
   46
Dividends, 209
Dog track, 46
Dollars, 209
Dynamic system, 217
Dynamism, 6

Earthquake, 12
Ecological
   integrity ethic, 225
   medicine, 10
   Society of America, 153
   slow-changing, 5
   - economic integration, 6
Ecology, 25
   agricultural, 64
   predictive , 64
Economic
   approach, 102, 104, 105,
   110
   criteria, 160
   determinism, 23
   growth, 220

paradigms, 4
productivity, 4
ECOPATH II, 203
Ecosystem
   activity, 9
   adaptability, 31
   adaptation, 37
   adaptive nature of , 246
   distress syndrome, 9,146,
   248
   flow analysis, 250
   flow networks, 196
   functional integrity, 49
   health, 3
   integrity, 25, 50, 59, 62, 63,
   65, 67, 68, 69, 124, 145, 150,
   152, 153, 180, 183, 226, 227,
   228, 229, 230, 231, 232, 233,
   235
   interconnections in 250
   level of activity and or
   ganization, 247
   net output, 215
   operational definitions of,
   239
   resilience, 9, 59, 129, 150,
   166, 178, 180, 230, 239, 243,
   246, 247, 249, 252, 253
   retrogression, 216
   stability, 7, 44, 49, 50, 124,
   137, 229, 233, 243, 246
   stress, 6, 135, 141, 147, 149,
   153, 161, 163, 243, 247, 248,
   252
   succession, 7, 45, 137, 171,
   180, 191, 193, 245
   succession-to-climax, 45
   theoretical , 46
   threshold, 158
   threshold criteria, 157, 162,
   167

Efficiency, 101, 107, 111
Electricity generating station, 216
Element cycling, 165
EMAP program, 252, 253
Employment, 207
Endangered species, 176
    Act, 226
Endpoint, 7, 191
Energy, 190, 200
    flow, 195, 231, 247
    intensities, 251
Entity
    orientation, 98, 110
    oriented, 109, 118
Entropy, 218
    high, 67
    low, 67
Environment
    abiotic, 31
Environmental
    damage, 212
    fascism, 30
    management, 15, 25
    policies, 15
    Protection Agency, 8
Equilibrium, 26, 158
Ethnocentrism, 44, 48
European, 45
Eutrophication, 67, 194, 249
Evolution, 86
Evolutionary, 45
    authenticity, 68
Exotic species, 146
Exotics, 12
Experimentalism, 36
Export, 209
Exported goods, 209
Exposure-response curves, 166
Externalities, 66, 208

Extinctions, 48
Farm, 50
Faust, 75
Faustian, 76, 77, 79, 81, 82, 91
    world, 77
Federal requirements, 161
Federal Reserve, 211
Finn index, 200
Fishery
    commercial, 11
    put-and-take, 63
Folk medicine, 125
Food chains, 13
Food web, 201
Forests, 220
    coniferous, 11
    national, 101
    tropical, 12
Free market, 46
Friedman, Milton , 111

Gaia, 109, 113
    theory, 28, 32
Genetic viability, 185
Genotype, 88
Geological, 45
Geologists, 62
Gifford Pinchot, 101, 224
Gleasonian individualistic paradigm, 46
Global climate change, 179, 223, 252
Global warming, 4, 46
GNP, 209
Goethe, 75
Golden trout, 54
Goods and services, 209
Grazing rights, 101
Great Lakes, 12
Great Slave Lake, 11

Great Smoky Mountains
National Park, 172, 177, 186
Greenhouse effect, 67
Gross national product, 209

Harvesting, 13
Health, 25
    care, 212
    index (HI), 248
Heisenberg Principle, 228
Heraclitus, 73
Herbivores, 214
Hetch Hetchy Valley, 174
Hierarchy, 6, 59, 65
Hierarchical
    approach, 4
    complexity, 30, 37
Holism, 29
Holistic, 50, 53, 80, 106, 110,
    153
    management, 33, 245, 246
Holocene, 45
Homemaker, 212
Homeostasis, 65, 112, 179,
    183, 187, 239, 243
Homeostatic, 177
Homo sapiens, 44
Human-health paradigm, 223
Hurricane, 12
Hydrological cycles, 177
Hypereutrophication, 13

Industrial fermentation
systems, 63
Ignorance, 83, 90, 92
    closed , 83
    communal, 85
    personal, 85
    reducible, 84, 91
    through complexity, 85
    through novelty, 85

Imported goods, 209
Income
    distribution, 103
    inequities, 212
Incommensurability, 104
Index
    of Biotic Integrity, 7, 164,
    217, 233, 234
    of interdependence, 251
    of system health, 253
Indicator species, 148
Indicators, 216
    short-term , 219
Individualistic paradigm, 45
Indonesia, 101
Industrial and postindustrial
societies, 5
Inflation, 210
Input/output (I-O) analysis,
    192, 250
Insect pest loads, 147
Institutional approaches, 115,
    118
    processes, 97
Instrumental good, 58
    value, 58, 59, 65, 66, 67, 69
    valuing, 43
Integration, 7
Intentional being, 27
Interbank loan rate, 211
Interest payments, 209
International Biosphere
Program., 153
Interrelatedness, 37
Intrinsic value, 59, 65, 67, 69,
    110, 112
Inventory of goods, 209
Inventory stocks, 209
Irreducible ignorance, 84, 91

Kaibab plateau, 126

Kant, 108, 109, 110
Keynes, 76
Keystone species, 170, 175
Kissimmee River, 79
Kudzu, 180

Labor, 207
Lake District in England, 68
Lake Erie, 136
Lake Okeechobee, 79
Land ethic, 175, 225
    Leopold's , 23
Land health, 47, 48, 49, 50, 54,
    124
Laplace's demon, 78
Laurentian Great Lakes, 148,
    232
Legal criteria, 160
Legionnaire's disease, 128
Legislation, 213
Leopold, Aldo, 25, 26, 30,
    175, 176
Life support systems, 223,
    225
Locus classicus, 47
Logical positivists, 111
Logos, 73, 74, 81
Lorenz, 88
Ludwig von Bertalanffy, 145

Macrobenthic Biotic Index,
    164
Macropolicy, 210
Manatees, 184
Manufacturing industries,
    220
Mass-balance models, 250
Maximum productivity, 65
Medical practitioners, 10
Medicine, 10
    preventive , 13

salutogenic, 149
Membranes, 29, 32, 34
Metallurgist, 62
Meteorological, 45
Midwest, 49
Mineral King Valley, 107
Minimalist systems ap-
    proach, 15, 32
Mixed-units, 250
Models
    input/output, 251
    modeling, 7
Monetary units, 210
Moral good, 58
Moses, 31
Muir, John, 173, 225
Multidimensional, 16
Multidisciplinary,3
Multimedia Environmental
Goals (MEG), 167
Multiscale, 240, 248, 251
Multispecies risk, 166

National Bureau of
Economics, 228
National output, 209
National Park Service, 176,
    184
National Parks Act, 232
National Program Guidance
for Biological Criteria, 232
Native American, 182
Natural balance, 171
Natural past, 182, 185, 187
Neocatastrophism, 46
Net output, 209
Network analysis, 14, 250,
    253
Network ascendency, 190,
    192
New Deal, 172

New paradigm, 37
New Physicists, 16
Newton, 31
Newtonian dynamics, 27
    mechanics, 29
Niche specialization, 8
Nitrogen, 216
NOAA, 8
Nonequilibrium, 141
    theories, 139
Noninstrumental valuing, 43
Nonreductionistic paradigm,
    29
Northern hardwoods, 11
Novelty, 90, 92
Nuclear power, 79
Nutrient, 216
    capital, 146
    cycles, 171
    cycling, 148, 165, 231
    retention, 147

Oak/pine forest, 11
Objective norms, 43, 47
Odum, 148, 192, 200, 247
Oikos, 73
Oil slick, 46
Old-growth forest, 46, 225
Open ignorance, 84, 85
Open system, 31, 72
Openness, 72, 81, 82, 86, 91,
    92
Opportunistic fishing, 11
Opportunity cost, 110
Optimal allocation, 66
Optimal scale, 66
Optimization models, 211
Organicism, 15, 27, 28, 29, 30,
    32
Organismic conception of
health, 129

theory, 135, 137,   139
Organization, 9, 247, 248, 249,
    250, 253
Organizational structure and
scale, 34
Overenrichment, 13
Ownership of capital, 212
Oxidants, 11
Ozone
    hole, 163
    layer, 67, 146

Paradigm shift, 139
Paradigmatic approach, 24
Pareto, 103, 111
Patient, 10
Per capita, 213
Perturbation, 245
Phenotypes, 88
Phenotypic and genotypic
diversity, 165
Phosphorus, 216
Piedmont, 185
Pipeline construction, 12
Plato, 31, 43, 47, 74, 80
Plimsoll line, 66
Pluralistic approach, 5
Policymaking, 209, 211
Political criteria, 160
Pollutant discharge, 8
Pollutants, 212
Population density, 163
Predictability, 9
Predicted Index of Biotic
Integrity (PIBI), 164
Prehuman times, 208
Prescriptive science, 17
Preservation ethic, 225
Price control, 211
Primary productivity, 11, 12,
    148, 216

Principal stocks, 208
Producers, 214
Production, 5
Profits, 207, 209
Prognosis, 10
Progressive differentiation, 145
    integration, 145

Rain forest, 102, 103, 110
Rate of growth, 207
Recruitment, 231
Reductionism, 23, 29
Reductionist management approach, 125
Reductionistic paradigm, 24, 27
Redwoods National Park, 177
Renaissance, 27, 29
Renewable resources, 11
Reproduction, 231
Resolution, 9
Resources
    conservation ethic, 224
    for the Future (RFF), 66
    management, 30
    natural, 207
Restoration, 183, 184, 187
    ecology, 136
Rio Gavilan, 26

Salutogenic perspective, 153
Scale, 9, 32, 34, 35, 36, 180, 182, 187, 192
    and degree of impact, 34
    and perspective, 33
    problems of scale, 36
    questions of scale, 35, 66
    scaling, 252

Science Advisory Board (SAB), 223, 240
Scientific criteria, 160, 163, 164
Scope for ascendency, 195, 197
Scope for growth, 195, 247
Scope index, 190
Sea level rise, 252
Sediment runoff, 8
Self-
    integrative, 4
    maintenance, 52
    organizing systems, 24
    recreating, 51
    regeneration, 52
    renewal, 47, 51
Shannon-Weaver index, 193
Shopping mall, 46
Simulation modeling, 14
    models, 252
Single unit of measure, 209
Size distribution, 11, 12, 216
Skinner, B. F. , 62
Snail darter, 107, 108
SO_2, 11, 220
Social
    criteria, 160
    factors, 162
    values, 17
    welfare policy, 210
Socrates, 42
Soil Conservation Service, 186
Soil erosion, 48
Spaceship Earth, 213
Species, 7
    composition, 147, 178
    diversity, 11, 12, 146, 150
    richness, 233
Stream water quality, 8

Stress index, 8
Study Design Assurance
analysis, 160
Subalpine lakes, 12
Sunlight, 216
    energy, 216
Supply rates, 207
Supraorganism, 30, 37
Surprise, 83, 85, 86, 90
Sustainability, 4, 25, 112, 117,
    229, 240, 248
Symptoms., 10
System connectivity models,
    165
Systematic organization, 26
Systemic creativity, 24
Systems analysis, 29
Systems science, 153
Systems theory, 28

Taxes, 209
Taxonomy, 153
Taxonomy of ecosystem ills,
    151
    of pathologies, 151
Technical progress, 207
Technological growth, 5
Teleology, 27
Tellico Dam, 107
Telos, 15
Temporal perspective, 33
Therapeutic nihilism, 53, 125,
Thermodynamics, 233
Three Mile Island, 141
Threshold criteria, 160, 164,
    166
Tidal marshes, 216
Total diversity, 26
Toxic
    mine tailings, 12
    waste dumping, 252

Trophic
    chain, 201
    dynamics, 231
    functioning, 190
    rank, 201

U.S. Fish and Wildlife
Service, 184
UCLA, 112
Uniformitarianism, 46
University of Wisconsin–
Madison Arboretum, 176, 181
Unstable state, 208
Urban sprawl, 46

Validation of treatments, 151
Valuation, 99, 103, 104, 105,
    107, 108, 110, 112, 113, 115,
    118, 119
Value, 5, 97, 108, 112, 210
Value added, 210
Values, 127, 241
Vigor, 247, 253
Vital signs., 10
Vitalism, 30

Wages, 209
Water Quality Act, 231
Waterways, 5
Western Hemisphere, 45
Wetlands, 46, 149
Whole-ecosystem
management, 35
Wilderness, 44, 68, 184
    idea, 45, 46
    myth, 53
W ork force, 212
Yellowstone, 36, 68
Yosemite National Park, 173

Contributors

SUSAN BRATTON, Department of Philosophy and Religion Studies, University of North Texas, Denton, Texas 76203. Susan has a Ph.D. in ecology from Cornell University and a graduate certificate in environmental ethics from the University of Georgia. She has served as the coordinator of the Uplands Field Research Laboratory in the Great Smoky Mountains National Park and of the U.S. National Park Service Cooperative at the University of Georgia. Her previous work includes two books: *Six Billion and More: Christian Ethics and Human Population Regulation* and *Christianity Wilderness and Wildlife: The Original Desert Solitaire* and a number of articles in journals such as *Environmental Ethics*. In addition, she has published widely in the fields of conservation biology, restoration ecology and park and wilderness management.

J. BAIRD CALLICOTT, Department of Philosophy, University of Wisconsin-Stevens Point, Stevens Point, WI 54481. Baird is Professor of Philosophy and Natural Resources at the University of Wisconsin-Stevens Point. One of the founders of the field of environmental ethics, he is author of *In Defense of the Land Ethic* and editor of *Companion to A Sand County Almanac*.

ROBERT COSTANZA, Chesapeake Biological Laboratory, University of Maryland, Solomons, MD 20688. In addition to serving as director of the newly formed Maryland International Institute for Ecological Economics, Bob is also a full Professor at the University of Maryland. He is editor of the book, *Ecological Economics: The Science and Management of Sustainability* (Columbia University Press 1991), and is co-founder (with Herman Daly) and chief editor of the journal *Ecological Economics*.

DAVID K. COX, College of Veterinary Medicine, University of Illinois at Urbana-Champaign, 2001 South Lincoln, Urbana, IL 61801. David is an Assistant Professor at the University of Illinois, an environmental lawyer and Vice-chair of the American Bar Association General Practice Environmental Law Committee. His research includes monitoring, analyzing, and forecasting environmental legislation; abstracting and computerization of environmental statutes; and development of policy and legal requirements for establishing ecosystem threshold criteria.

DAVID EHRENFELD, Department of Natural Resources, Rutgers University, New Brunswick, NJ 08903. David received his M.D. from Harvard Medical School and his Ph.D. (in zoology) from the University of Florida. He is professor of biology at Rutgers University, editor of *Conservation Biology*, and author of several books including *The Arrogance of Humanism* and a forthcoming book on society and the environment entitled *Beginning Again; People and Nature in the New Millennium*, to be published by Oxford University Press.

MALTE FABER, Alfred-Weber-Institut, University of Heidelberg, Grabengasse 14, D-6900, Heidelberg, Germany. Malte has been Professor of Economic Theory at the University of Heidelberg since 1973. He has published widely in capital theory, public choice, and the role of the entropy concept in environmental economics, with applications to water and waste management. He has served as an adviser on environmental matters to the federal and state governments of the Federal Republic of Germany.

BRUCE HANNON, National Center for Supercomputing Applications and the Department of Geography, University of Illinois, Urbana, IL 61801. Bruce is professor at the National Center for Supercomputing Applications and Jubilee Professor of Literal Arts and Sciences at the University of Illinois. He has been involved in ecological and economic modeling for the last 25 years.

EUGENE HARGROVE, Department of Philosophy and Religion Studies, University of North Texas, Denton, Texas 76203. Gene, chair of the Department of Philosophy and Religion Studies, is the editor of *Environmental Ethics: An interdisciplinary journal dedicated to the philosophical aspects of environmental problems* and is the author of *Foundations of Environmental Ethics* (Prentice-Hall, 1989).

BENJAMIN D. HASKELL, Maryland Sea Grant College, 1123 Taliaferro Hall, University of Maryland, College Park, MD 20742. Ben is a graduate student in the Marine-Estuarine-Environmental

Sciences Program at the University of Maryland and a Dean John A. Knauss Marine Policy Fellow with NOAA's Sanctuaries and Reserves Division. He is studying coastal zone management issues.

JAMES R. KARR, Institute for Environmental Studies, University of Washington, Seattle, WA 98195. Jim is Director of the Institute for Environmental Studies and Professor in Zoology and Fisheries at the University of Washington. His current research interests include water resources, landscape ecology, avian demography, and conservation biology, with special interest in the use of ecological science to inform environmental and natural resource decisions.

REINER MANSTETTEN, Alfred-Weber-Institut, University of Heidelberg, Grabengasse 14, D-6900 Heidelberg, Germany. Reiner is an assistant in economic theory at the Department of Economics. He holds an M.A. in philosophy and has just submitted his doctoral thesis to the Department of Philosophy. He has published jointly with Malte Faber and John Proops on the philosophical foundations of ecological economics.

BRYAN G. NORTON, School of Public Policy, Georgia Institute of Technology, Atlanta, GA 30332. Bryan studied philosophy of science at the University of Michigan; he is currently Professor of Philosophy of Science and Technology at Georgia Institute of Technology. He is the author of *Linguistic Frameworks and Ontology, Why Preserve Natural Variety?* and *Toward Unity Among Environmentalists*; he is editor of the *Preservation of Species* and is author of numerous journal articles in environmental ethics and environmental policy.

TALBOT PAGE, Economics Department, Brown University, Providence, Rhode 02912. Toby is Professor of Economics at Brown University. He has written on normative economics, resource economics, environmental regulation, policy analysis, decision theory, game theory, and the epidemiology of drinking water. His wife is a painter and they have two chilldren.

JOHN PROOPS, Department of Economics, Keele University, Staffordshire ST5 5BG, United Kingdom. John is a Senior Lecturer in Economics at Keele University. His main research interests are in the conceptual foundations of ecological economics, the applications of capital theory to long-run economy-environment interactions, and the applications of input-output analysis to environmental issues.

DAVID J. RAPPORT, Institute for Research on Environment and Economy, University of Ottawa, 5 Calixa Lavallee, Ottawa, Ontario K1N 6N5 Canada. David is Adjunct Professor of Biology at the University of Ottawa and is a senior research scientist with Statistics Canada. He co-authored Canada's first State of the Environment Report.

MARK SAGOFF, Institute for Philosophy and Public Policy, University of Maryland, College Park 20742. Mark is Director and Senior Research Scholar at the Institute. He is the author of *The Economy of the Earth: Philosophy, Law, and the Environment* (Cambridge University Press).

DAVID J. SCHAEFFER, Deparment of Veterinary Biosciences, University of Illinois at Urbana-Champaign, 2001 South Lincoln, Urbana, IL 61801. David is Research Professor in Ecosystem Toxicology at the University of Illinois and was Science Advisor to the Illinois Environmental Protection Agency for 14 years. His research interests include statistical modeling, development of bioassay batteries for ecotoxicology, and the conceptual and theoretical aspects of ecosystem medicine.

ROBERT E. ULANOWICZ, Chesapeake Biological Laboratory, University of Maryland, Solomons, MD 20688. Bob is a Professor of Ecosystems Research at the Chesapeake Biological Laboratory, where he has worked for 22 years. His research interests include the analysis of ecological flow networks, information theory in ecology, the notion of causality in scientific discourse, and the ecology of Florida sinkholes.

ALSO AVAILABLE FROM ISLAND PRESS

Balancing on the Brink of Extinction: The Endangered Species Act and Lessons for the Future
Edited by Kathryn A. Kohm

Better Trout Habitat: A Guide to Stream Restoration and Management
By Christopher J. Hunter

Beyond 40 Percent: Record-Setting Recycling and Composting Programs
By The Institute for Local Self-Reliance

Coastal Alert: Ecosystems, Energy, and Offshore Oil Drilling
By Dwight Holing

The Complete Guide to Environmental Careers
By The CEIP Fund

Crossing the Next Meridian: Land, Water, and the Future of the West
By Charles F. Wilkinson

Death in the Marsh
By Tom Harris

The Energy-Environment Connection
Edited by Jack. M. Hollander

Farming in Nature's Image
By Judith Soule and Jon Piper

Ghost Bears: Exploring the Biodiversity Crisis
By R. Edward Grumbine

The Global Citizen
By Donella Meadows

Green at Work: Making Your Business Career Work for the Environment
By Susan Cohn

Healthy Homes, Healthy Kids
By Joyce Schoemaker and Charity Vitale

Holistic Resource Management
By Allan Savory

The Island Press Bibliography of Environmental Literature
By The Yale School of Forestry and Environmental Studies

Last Animals at the Zoo: How Mass Extinction Can Be Stopped
By Colin Tudge

Learning to Listen to the Land
Edited by Bill Willers

Lessons from Nature: Learning to Live Sustainably on the Earth
By Daniel D. Chiras

The Living Ocean: Understanding and Protecting Marine Biodiversity
By Boyce Thorne-Miller and John G. Catena

Making Things Happen
By Joan Wolfe

Nature Tourism: Managing for the Environment
Edited by Tensie Whelan

Not by Timber Alone
By Theodore Panayotou and Peter S. Ashton

Our Country, The Planet: Forging a Partnership for Survival
By Shridath Ramphal

Overtapped Oasis: Reform or Revolution for Western Water
By Marc Reisner and Sarah Bates

Population, Technology, and Lifestyle: The Transition to Sustainability
Edited by Robert Goodland, Herman E. Daly, and Salah El Serafy

Rain Forest in Your Kitchen: The Hidden Connection Between Extinction and Your Supermarket
By Martin Teitel

The Snake River: Window to the West
By Tim Palmer

Spirit of Place
By Frederick Turner

Taking Out the Trash: A No-Nonsense Guide to Recycling
By Jennifer Carless

Turning the Tide: Saving the Chesapeake Bay
By Tom Horton and William M. Eichbaum

Visions upon the Land: Man and Nature upon the Western Range
By Karl Hess, Jr.

The Wilderness Condition
Edited by Max Oelschlaeger

For a complete catalog of Island Press publications, please write:
Island Press, Box 7, Covelo, CA 95428, or call: 800-828-1302